United States Nuclear Regulatory Commission

Protecting People and the Environment

NUREG/CR-7159
PNNL-14561

I0482743

Reliability of Ultrasonic In-Service Inspection of Welds in Reactor Internals of Boiling Water Reactors

Office of Nuclear Regulatory Research

AVAILABILITY OF REFERENCE MATERIALS
IN NRC PUBLICATIONS

NRC Reference Material

As of November 1999, you may electronically access NUREG-series publications and other NRC records at NRC's Public Electronic Reading Room at http://www.nrc.gov/reading-rm.html. Publicly released records include, to name a few, NUREG-series publications; *Federal Register* notices; applicant, licensee, and vendor documents and correspondence; NRC correspondence and internal memoranda; bulletins and information notices; inspection and investigative reports; licensee event reports; and Commission papers and their attachments.

NRC publications in the NUREG series, NRC regulations, and Title 10, "Energy," in the *Code of Federal Regulations* may also be purchased from one of these two sources.
1. The Superintendent of Documents
    U.S. Government Printing Office Mail Stop SSOP
    Washington, DC 20402–0001
    Internet: bookstore.gpo.gov
    Telephone: 202-512-1800
    Fax: 202-512-2250
2. The National Technical Information Service
    Springfield, VA 22161–0002
    www.ntis.gov
    1–800–553–6847 or, locally, 703–605–6000

A single copy of each NRC draft report for comment is available free, to the extent of supply, upon written request as follows:
Address: U.S. Nuclear Regulatory Commission
            Office of Administration
            Publications Branch
            Washington, DC 20555-0001
E-mail: DISTRIBUTION.RESOURCE@NRC.GOV
Facsimile: 301–415–2289

Some publications in the NUREG series that are posted at NRC's Web site address http://www.nrc.gov/reading-rm/doc-collections/nuregs are updated periodically and may differ from the last printed version. Although references to material found on a Web site bear the date the material was accessed, the material available on the date cited may subsequently be removed from the site.

Non-NRC Reference Material

Documents available from public and special technical libraries include all open literature items, such as books, journal articles, transactions, *Federal Register* notices, Federal and State legislation, and congressional reports. Such documents as theses, dissertations, foreign reports and translations, and non-NRC conference proceedings may be purchased from their sponsoring organization.

Copies of industry codes and standards used in a substantive manner in the NRC regulatory process are maintained at—
    The NRC Technical Library
    Two White Flint North
    11545 Rockville Pike
    Rockville, MD 20852–2738

These standards are available in the library for reference use by the public. Codes and standards are usually copyrighted and may be purchased from the originating organization or, if they are American National Standards, from—
    American National Standards Institute
    11 West 42nd Street
    New York, NY 10036–8002
    www.ansi.org
    212–642–4900

Legally binding regulatory requirements are stated only in laws; NRC regulations; licenses, including technical specifications; or orders, not in NUREG-series publications. The views expressed in contractor-prepared publications in this series are not necessarily those of the NRC.

The NUREG series comprises (1) technical and administrative reports and books prepared by the staff (NUREG–XXXX) or agency contractors (NUREG/CR–XXXX), (2) proceedings of conferences (NUREG/CP–XXXX), (3) reports resulting from international agreements (NUREG/IA–XXXX), (4) brochures (NUREG/BR–XXXX), and (5) compilations of legal decisions and orders of the Commission and Atomic and Safety Licensing Boards and of Directors' decisions under Section 2.206 of NRC's regulations (NUREG–0750).

NUREG/CR-7159
PNNL-14561

United States Nuclear Regulatory Commission

Protecting People and the Environment

Reliability of Ultrasonic In-Service Inspection of Welds in Reactor Internals of Boiling Water Reactors

Manuscript Completed: November 2012
Date Published: April 2013

Prepared by:
G. J. Schuster, S. L. Crawford, A. A. Diaz,
P. G. Heasler, and S. R. Doctor

Pacific Northwest National Laboratory
P. O. Box 999
Richland, WA 99352

D. A. Jackson and W. E. Norris, NRC Project Managers

NRC Job Codes Y6604 and N6398

Office of Nuclear Regulatory Research

ABSTRACT

Instances of stress corrosion cracking in reactor pressure vessel internal components have been found, especially in the boiling water reactor (BWR) core shroud. Results from in-service inspection are an important aspect of integrity evaluations. One of the major goals of the work described in this report is to quantify the crack detection reliability and sizing error of ultrasonic inspection methods for reactor internals. A mockup is described, along with its application to the assessment on nondestructive evaluation reliability for in-service inspection of reactor internals. The mockup of a BWR core shroud includes cracks in some of the 40 welded assemblies selected to represent field conditions. The selected material and geometry include most conditions and alloys used in the core shroud and its support structure. This report provides an overview of the work being performed and focuses on the parametric study results relating ultrasonic response and its variance to inspection effectiveness. Results of a blind test of an in-service inspection vendor's phased array technique are also provided in this report.

PAPERWORK REDUCTION ACT STATEMENT

PUBLIC PROTECTION NOTIFICATION

FOREWORD

The Nuclear Regulatory Commission (NRC) conducted research at Pacific Northwest National Laboratory (PNNL) to assess the density and distribution of flaws in reactor pressure vessel (RPV) welds that result from fabrication and repair processes. The research was initiated because analyses have shown that vessel behavior is sensitive to flaw location, type, size, orientation, and other flaw characteristics. Accurate estimates of flaw density and distribution are required as input to the computer codes that are used in performing structural integrity assessments. Research results were published in NUREG/CR-6945, *Fabrication Flaw Density and Distribution in Repairs to Reactor Pressure Vessel and Piping Welds.*

The subject report describes earlier work to quantify the crack detection reliability and sizing error of ultrasonic inspection methods for reactor internals. A parametric study was used to divide those inspections where performance was considered acceptable from those where inspection was considered difficult. This was accomplished by quantifying crack responses and attendant noise sources to show that reliable detection and accurate sizing were related to a high signal-to-systematic-noise ratio. A blind test of vendor performance was used to measure detection and sizing statistics for the difficult cases.

In publishing this report, the NRC's purpose is to make the research results publicly available because the vendor's phased array in-service inspection was conducted under blind test conditions, and this information will be a valuable source of information for future performance demonstration assessment. In addition, the results of the study provide insights related to reliable detection and accurate sizing.

CONTENTS

FIGURES

TABLES

EXECUTIVE SUMMARY

As a part of the Nuclear Regulatory Commission (NRC) Office of Nuclear Regulatory Research activities on evaluating the reliability of nondestructive examinations, the Pacific Northwest National Laboratory has measured the reliability of in-service inspection of boiling water reactor (BWR) internals components for stress corrosion cracking of the weld's heat-affected zones. The primary objective of this study was to identify areas or conditions in which inspection performance was relatively low. A mockup of a BWR core shroud was used to quantify inspection performance.

Inspection performance was quantified in this study by identifying the areas in which performance was high and then placing emphasis on the remaining difficult inspection cases. The identification of high-performance conditions was accomplished by use of a parametric study, which quantified crack responses and attendant noise sources to show that reliable detection and accurate sizing were related to a high signal-to-systematic-noise ratio. A blind test of vendor performance was used to measure detection and sizing statistics for the difficult cases.

Results of the parametric study showed that access to the cracked surface (the crack initiation surface) strongly affected detection and sizing. For the inspection of weld heat-affected zones, reliable detection of cracking and accurate sizing could be accomplished when an ultrasonic transducer could be placed on the cracked surface. Furthermore, reliable detection and accurate sizing could be achieved for the near-side access condition to cracks that are connected to the surface opposite from the surface used for scanning. The parametric study showed that for far-side access (through the weld) to cracks connected to the opposite surface, detection and sizing performance was low and the most difficult of inspection conditions.

A blind test of vendor performance for weld far-side access to heat-affected zone cracking also was conducted. The vendor used a phased array system to successfully detect 14 of 17 cracked grading units while calling cracks in 3 of 33 blank grading units. This detection performance would meet the intent of the requirements of an ASME Code Section XI, Appendix VIII, performance demonstration test designed to Supplement 2, "Qualification Requirements for Wrought Austenitic Piping Welds," because there were twice as many blank grading units as cracked grading units. The root mean square error for the blind depth-sizing test was 9.2 mm and would not pass.

A re-analysis of the signals in the vendor's phased array data revealed that no response was received from cracks less than 30% through-wall for far-side access. The absence of corner-trap responses explains, in part, the size dependence of crack detection. The origin of the signals in the phased array images was from the rough face of the cracks and from the crack tips. Sizing errors occurred because weak responses from crack tips were difficult to distinguish from strong responses from the crack face and from weld fabrication flaws.

Techniques are described for imaging improvements, such as reduction of noise variance and use of hybrid imaging systems. Application of new technology to the difficult case of far-side inspection of austenitic welds is needed. The industry has worked extremely hard on the far-side inspection issue for Appendix VIII Supplement 10 because field dissimilar metal welds limit most examinations to be conducted only from the far side. Some of the Supplement 10 solutions may provide improvements to the BWR reactor internals inspections.

ACKNOWLEDGMENTS

The authors thank Deborah A. Jackson (now at NRC NMSS) and Wallace E. Norris of NRC RES for their technical and managerial guidance along with the continued support to being this work to successful closure. We also thank J. Muscara formerly of NRC RES (but now retired) for starting this technical work and for his initial technical guidance.

The authors thank S. H. Bush of PNNL for assistance with the configuration of light water reactor internals and A. F. Pardini of PNNL for assistance in the parametric study. The authors thank R. E. Bowey of PNNL for assistance in the blind test of the phased array system. We thank E. Prickett and K. E. Hass of PNNL for preparing and editing this manuscript.

We thank the staff of the Electric Power Research Center NDE Center for hosting our activities during the blind test of the phased array system at their facility. The authors thank John Hayden of G.E. Nuclear Inspection Service for participating in a blind test of the PNNL reactor internal mockup using the G.E. phased array system.

ABBREVIATIONS AND ACRONYMS

ASME	American Society of Mechanical Engineers
BWR	boiling water reactor
BWROG	Boiling Water Reactors Owners Group
BWRVIP	Boiling Water Reactor Vessel and Internals Project
EPRI	Electric Power Research Institute
FCP	false call probability
FDF	flaw detection frequency
IGSCC	intergranular stress corrosion cracking
INPO	Institute of Nuclear Power Operations
ISI	in-service inspection
L	longitudinal
LPCI	low pressure coolant injection
MRR	mini-round robin
NDE	nondestructive evaluation
NDT	nondestructive testing
NRC	U.S. Nuclear Regulatory Commission
OC	operating characteristics
PDI	Performance Demonstration Initiative
PIRR	Piping Inspection Round Robin
PISC	Programme for the Inspection of Steel Components
PNNL	Pacific Northwest National Laboratory
POD	probability of detection
RMSE	root mean square error
RPV	reactor pressure vessel
SCC	stress corrosion cracking (cracks)
SE	safety evaluation
SG-CSS	Subgroup on In-Service Inspection of Core Support and Internal Structures
SH	shear
TFC	thermal fatigue crack
TOFD	time-of-flight diffraction
UT	ultrasonic testing
VT	visual testing
WSC	weld solidification crack

1 INTRODUCTION

As a part of the Nuclear Regulatory Commission (NRC) Office of Nuclear Regulatory Research program, "Evaluation of the Reliability of Non Destructive Examination Techniques," and the follow-on program, "Reliability of Nondestructive Examination for Nuclear Power Plant Inservice Inspection," the Pacific Northwest National Laboratory has assessed the reliability of in-service inspection of boiling water reactor (BWR) internals components for stress corrosion cracking of the weld's heat-affected zones. This report provides estimates of flaw detection capability and sizing accuracy for the ultrasonic testing (UT) methods that have been employed. Also, the report provides an assessment of component accessibility and describes its effect on inspection capability. This report includes UT data from a realistic mockup, data assessment, and findings.

The ability of nondestructive evaluation (NDE) to reliably detect and accurately size stress corrosion cracks in stainless steel piping has been studied in research programs for more than 20 years. Measurement of the performance of standard and advanced NDE practice has been made in round robins conducted by Pacific Northwest National Laboratory (PNNL) (Heasler et al. 1990; Heasler and Doctor 1996) and in the international Programme for the Inspection of Steel Components (Lemaitre et al. 1996). This report summarizes the results from these extensive databases to compare and contrast the inspection of piping to the inspection of reactor internals.

The purpose of this research was to determine the detection and sizing capabilities of ultrasonic inspections as they apply to in-service inspection (ISI) of reactor internals. This evaluation of NDE reliability was to be applicable to the techniques practiced in the field. Special attention was given to the identification of areas in which performance may be relatively low.

Figure 1-1 shows a conceptual diagram for the study of the UT-ISI reliability as it was applied to reactor internals. The activity at the center of the work was a parametric study of inspection performance. The goal of the parametric study was to identify areas in which performance was relatively low. For the cases of low performance, this study employed a mockup in a blind test to measure the performance of industry-deployed techniques that were thought to provide enhancements in the inspection. The performance in the best case has been combined with the estimates for the difficult cases to establish the NDE reliability of the ISI of reactor internals. This study also used failure history, especially crack types, and the physical properties of ISI methods to augment the understanding of reliability. At the bottom of Figure 1-1, the mockup of reactor internals provides the measurement environment for both the parametric study and the blind tests. The literature on NDE reliability measurement shows how a response model can be used in a parametric study to estimate detection and sizing performance.

Section 2 contains a description of BWR internals and intergranular stress corrosion cracking that occurs in the heat-affected zones of the welds. The BWR internals are presented as a way of introducing the inspection problem. The welds in the BWR core shroud are described, and selected other components are illustrated. The physical properties of stress corrosion cracks (SCCs) are given, especially those important to ultrasonic inspection.

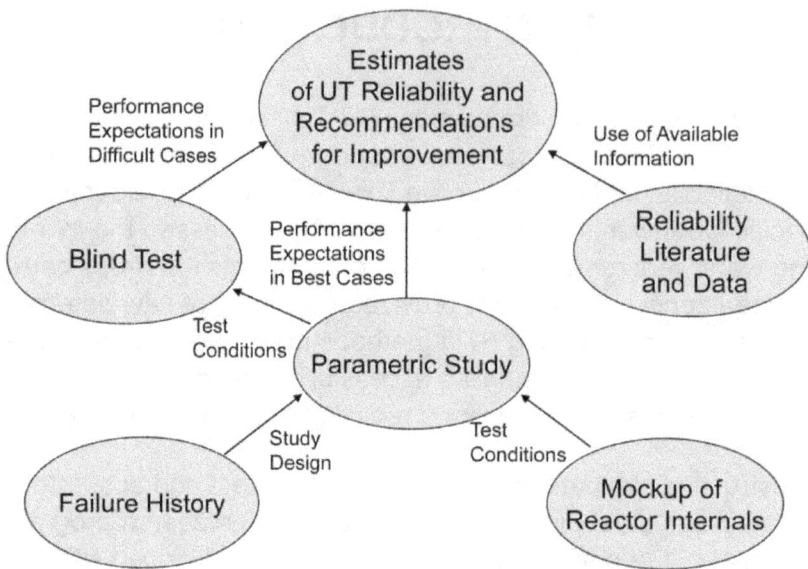

Figure 1-1   Conceptual Diagram for Study of UT-ISI Reliability as Applied to Reactor Internals

Section 3 contains a description of the ultrasonic testing methods applied to the in-service inspection of BWR internals. The ultrasonic modalities covered are 45° shear waves, high-angle compression waves, creeping waves, time of flight diffraction, and phased arrays. A brief description of mechanical deployment is provided for perspective on the application of NDE technologies.

Section 4 contains a description of the NDE reliability literature. Research results from PNNL piping round robins and the Programme for the Inspection of Steel Components (PISC) are reviewed in Section 4. The use of the relative operating characteristics diagram is explained and an example of its use is given from the literature. The relevance of responses from degradation and from noise sources in inspection reliability also are reviewed in this section.

Section 5 contains a description of the mockup built for this study. The geometrical conditions of the BWR core shroud that affect NDE reliability are given and included in the mockup design. The mockup was constructed for use in both a parametric study and a blind test. The results of the flaw acceptance tests are provided, and a list of the mockup assemblies is given.

Section 6 describes the parametric study. Weld near-side access to SCC is shown to provide a highly reliable inspection. The echo-dynamics of corner trap signals is reviewed. Inspection reliability for weld far-side access to cracks is shown. Data from cracks connected to the scanning surface is also reported.

Section 7 contains a description of a blind test for inspection performance through austenitic weld and the statistical results from it. Detection performance was measured using the grading unit approach specified for performance demonstration (American Society of Mechanical Engineers [ASME] Boiler and Pressure Vessel Code Section XI, Appendix VIII) (ASME 2004).

Depth sizing and length sizing results are given with a parametric analysis of the blind test data for the case of far-side access (through the weld).

Sections 8 and 9 present the conclusions and recommendations based on this work.

2 WELDS IN BWR INTERNALS

Components associated (both internally and externally) with the BWR reactor pressure vessel (RPV) vary widely in overall geometry, size, shape, and thickness. The configuration of BWR internals is documented in Shaw and McDonald (1993) and in BWRVIP-15 (1996). Figure 2-1 (permission to use this copyrighted material is granted by the American Nuclear Society) depicts the arrangement of the reactor internals used in the General Electric BWR model 4. The location of the core shroud in the internals assembly can be seen in Figure 2-1.

There are three regions—the lower, central, and upper regions—that make up the volume enclosed by the BWR pressure vessel. The lower region, below the core plate, contains the lower plenum components described in BWRVIP-47 (1997) and the core shroud support structure. The central region, bounded by the top guide and the core plate, is the area containing the fuel. The upper region is bounded by the shroud head and the top guide.

All three regions play an important part in the flow of energy from the reactor core to the turbine generators. Feedwater enters the vessel through the feedwater nozzle, is mixed with internal recirculating water, and flows into the annulus between the vessel wall and the core shroud. From the annulus, a portion of water leaves the vessel via the recirculating water outlet nozzle; the remainder enters the suction inlets of the jet pump assemblies. The external recirculation water is pumped through the recirculation loops and returns to the nozzle inlets of the jet pumps. All three streams join together in the pump diffuser and discharge into the plenum below the core. The water then flows up through the core where a portion of the liquid is evaporated by boiling into steam. This mixture of steam and water passes up through the steam separators and dryers. The separated water flows back down to the annulus for recirculation while the steam is sent to the turbine generator.

The core shroud provides vertical and lateral support for the top guide and core plate and provides a floodable volume for emergency cooling water. The core shroud is a stainless steel cylindrical assembly that supports the core plate, shroud head, and top guide both circumferentially and axially. The shroud head is manufactured of wrought austenitic stainless steel and closes off the core outlet, thereby forcing the steam to travel through the steam separators.

2.1 BWR Internals Weld Locations

Reactor internals are welded assemblies of wrought austenitic stainless steel, Alloy 600, and cast austenitic stainless steel. These welded assemblies include the core shroud, top guide, core plate, jet pump assembly, core spray internals, RPV attachments, shroud support assembly, liquid standby control, and lower plenum components.

Figure 2-2 illustrates the typical BWR shroud layout with major circumferential welds. The weld locations for typical General Electric shroud designs for BWR models 3, 4, 5, and 6 are shown. The core shroud is a cylindrical, stainless steel assembly surrounding the core and provides a barrier for separating upward flow through the core from downward flow in the annulus.

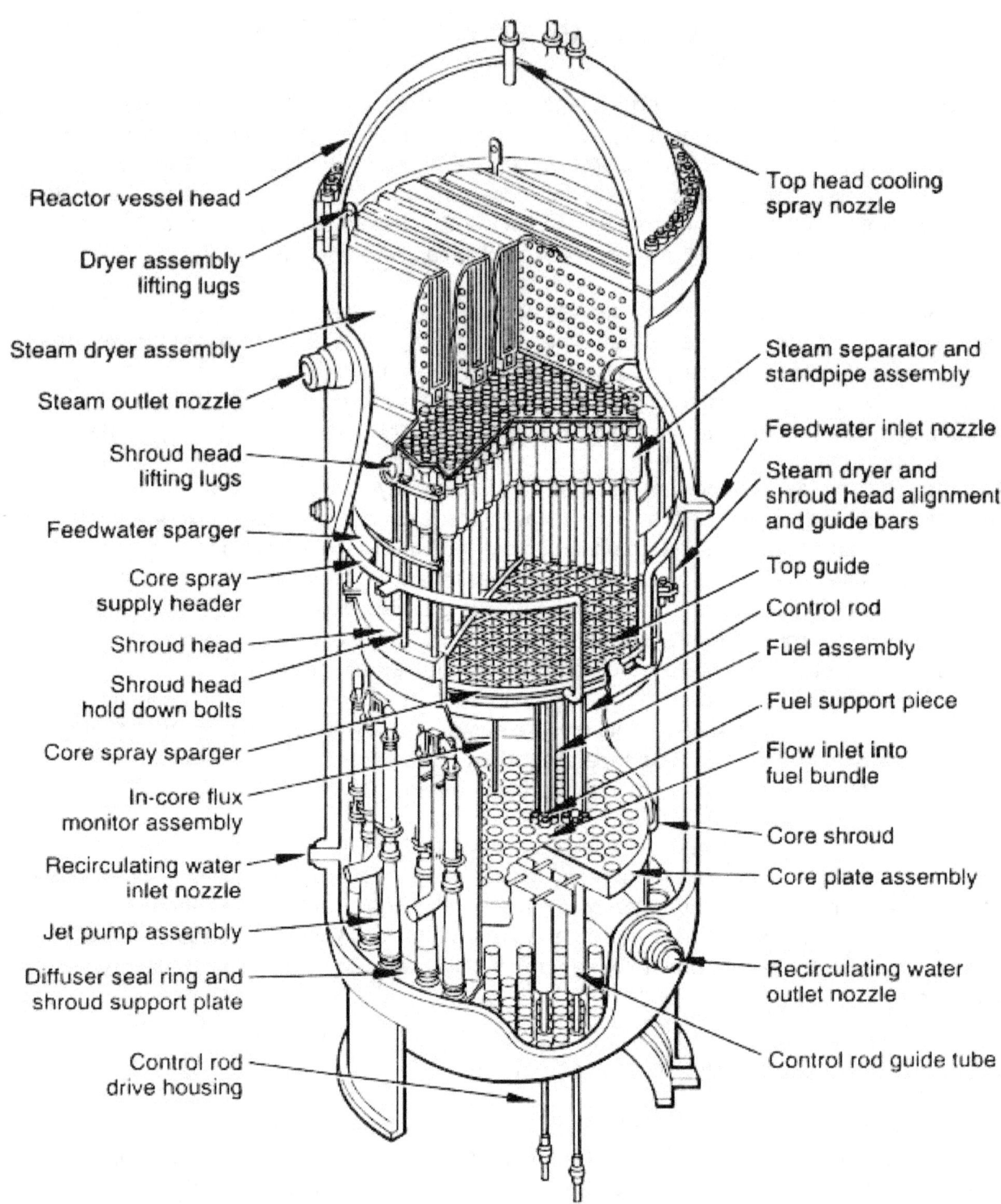

Reactor vessel head

Dryer assembly lifting lugs

Steam dryer assembly

Steam outlet nozzle

Shroud head lifting lugs

Feedwater sparger

Core spray supply header

Shroud head

Shroud head hold down bolts

Core spray sparger

In-core flux monitor assembly

Recirculating water inlet nozzle

Jet pump assembly

Diffuser seal ring and shroud support plate

Control rod drive housing

Top head cooling spray nozzle

Steam separator and standpipe assembly

Feedwater inlet nozzle

Steam dryer and shroud head alignment and guide bars

Top guide

Control rod

Fuel assembly

Fuel support piece

Flow inlet into fuel bundle

Core shroud

Core plate assembly

Recirculating water outlet nozzle

Control rod guide tube

Figure 2-1   Cutaway View of BWR-4 (Herrera and Stancavage 1988). Copyright 1988 by the American Nuclear Society (ANS), La Grange Park, Illinois; reprinted with permission.

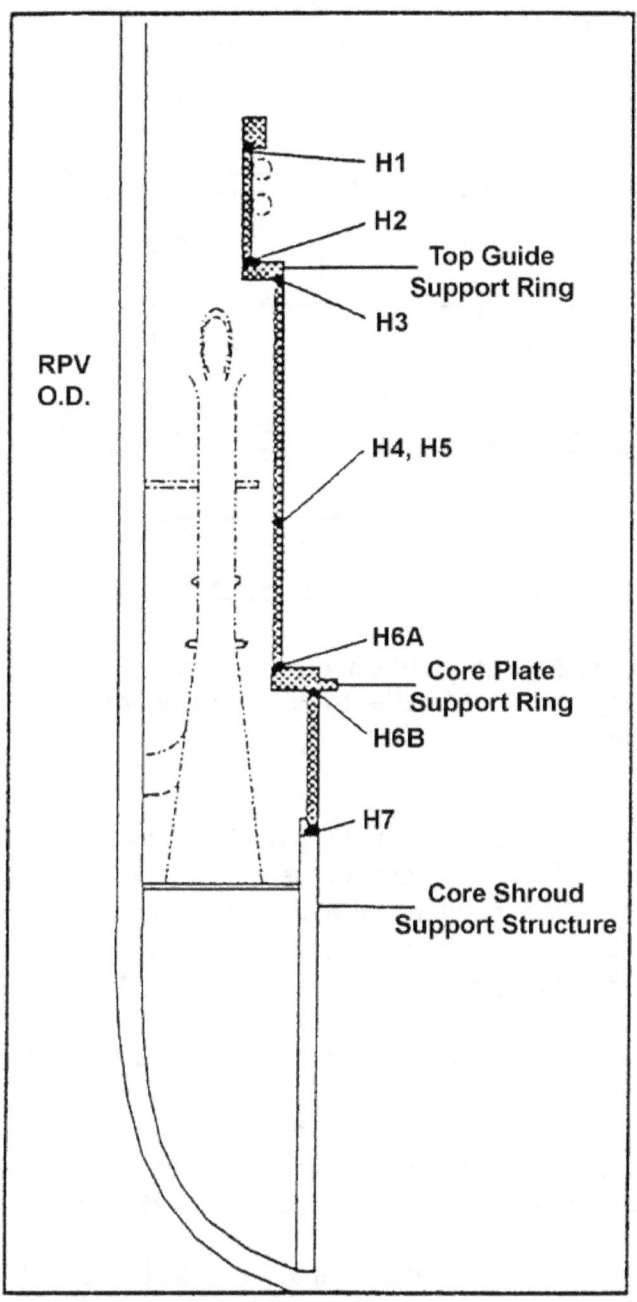

Figure 2-2   Cutaway View of Typical Core Shroud. Source: Baker and Van Hoomissen (1994), copyright © 1994 Electric Power Research Institute (EPRI), Inc.; reproduced with permission.

The core shroud is made up of several panels, two support rings (at the core plate and top guide elevations), and a third ring and steam dam at the shroud head. This entire assembly is held together by numerous vertical and circumferential welds (seams), as shown in Figure 2-3 welds (permission to use this copyrighted material is granted by the Electric Power Research Institute). In a typical BWR, there are seven or eight circumferential welds, and these welds are where cracking has been found.

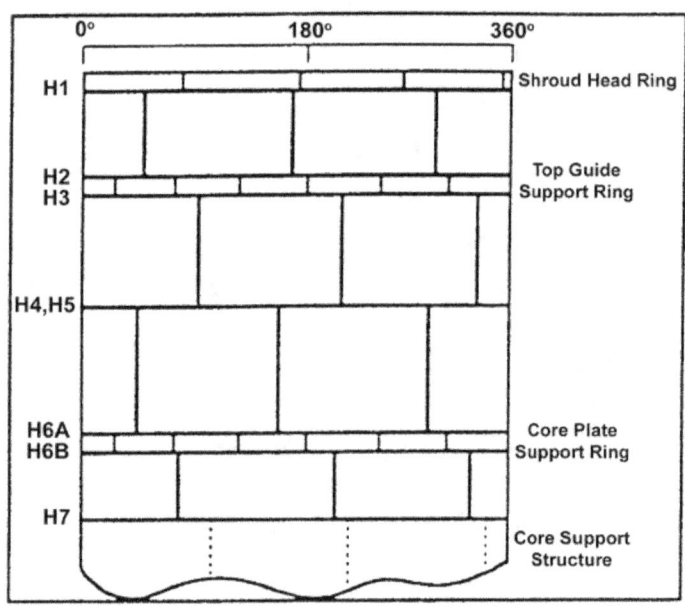

Figure 2-3   Typical BWR Shroud Weld Layout. Source: Baker and Van Hoomissen (1994), copyright © 1994 Electric Power Research Institute (EPRI, Inc.; reproduced with permission.

The locations of the welds associated with the top guide are illustrated in BWRVIP-26 (1996). The various wedges and hold-down devices for the top guide also are illustrated there. The top guide is bolted on a rim near the top of the core shroud and provides lateral support for the upper end of the fuel assemblies, neutron sources, and neutron monitoring instruments. Stress corrosion cracking has been found in the top guide assembly (BWRVIP-26 1996).

The core plate assembly is a stainless steel circular plate, oriented horizontally, with axial stiffener plates beneath it. This plate provides support for the peripheral fuel bundles and is bolted to the support ledge in the bottom area of the core shroud. The various BWR core plate welds are illustrated clearly in BWRVIP-25 (1996). Stress corrosion cracking has been found in the heat-affected zone of welds in the core plate (BWRVIP-25 1996).

The jet pumps play a critical role in the recirculation process of the BWR RPV. The flow is throttled by a flow control valve on the discharge side of the recirculation pump. A cutaway view of a typical jet pump assembly is shown in Figure 2-4. The position and orientation of the jet pump is shown relative to the shroud plate and the RPV recirculating inlet nozzle.

More jet pump assembly details including welds, screws, shims, brackets, elbows, springs, and collars can be found in great detail with the potential failure locations in BWRVIP-41 (1997).

In a BWR, a high-pressure core spray system is provided to make up water lost in any break or rupture of the primary system and to prevent the core from overheating. Water is sprayed directly onto the core using the core spray sparger. A low-pressure core spray system (with associated spray sparger) sprays water directly onto the fuel elements and is triggered during high containment pressure conditions or low reactor water level conditions. A clear view of

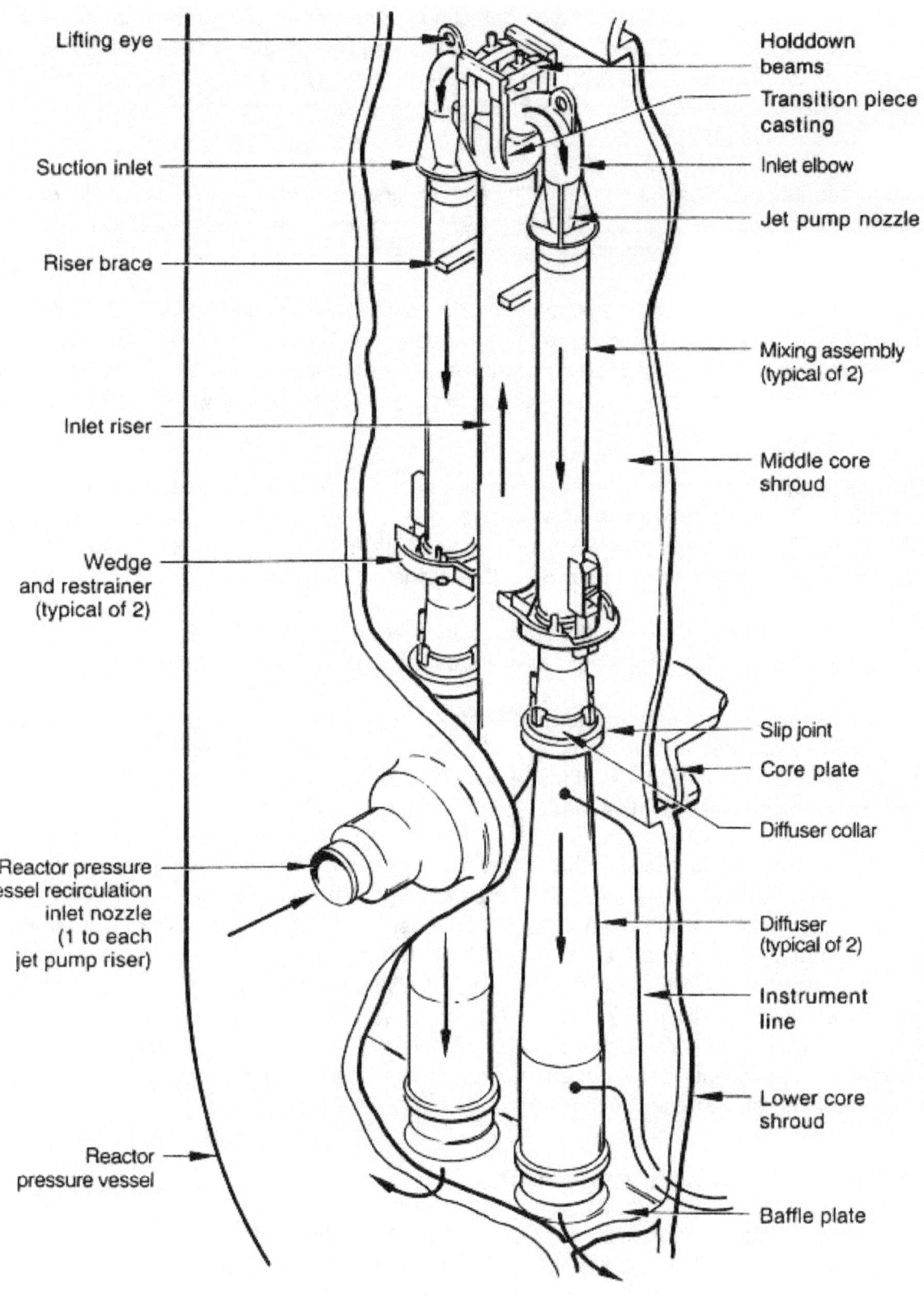

Figure 2-4   Cutaway View of Jet Pump Assembly (USNRC Technical Training Center)

the typical core spray piping configuration depicting support brackets, the shroud, and shroud attachment is shown in BWRVIP-15 (1996). The core spray sparger is part of the BWR emergency cooling system. Cracking occurs in the core spray spargers, in particular the T-box/vent hole configuration. The inspection guidelines for the core spray internals can be found in BWRVIP-18 (1996).

The configuration, degradation susceptibility factors, and potential failure location for vessel inside attachments is given in BWPVIP-48 (1998). Mechanical damage and SCC have been found in the attachments (BWRVIP-48 1998).

The welds in the shroud support assembly for BWRs can be found in BWRVIP-15 (1996). The configuration of the shroud support assembly varies by BWR plant type, and the variations are described there. Ultrasonic inspections of the H8 and H9 welds have been conducted at more than 20 BWRs, and minor indications (those with no significant depth) were found (BWRVIP-38, 1997).

The shroud support assembly has access holes that are used as manways during plant construction. These holes are sealed with cover plates made of Alloy 600 and are about 6.4-cm (2.5-in.) thick. The covers are then welded in place with Alloy 82 or 182 weld metal. Alloy 182 material is known to be susceptible to SCC (MacDonald 1994).

The welds associated with the liquid standby control are given in BWRVIP-15 (1996). The inspection and flaw evaluation guidelines are presented in BWRVIP-27 (1997).

The lower plenum components include the control rod guide tube, control rod housing, stub tube and orificed fuel support, in-core housing and support hardware, and the in-core flux monitor dry tube assembly. The configuration of these components, the failure history, and inspection guidelines are given in BWRVIP-47 (1997).

The BWRVIP reports addressed in this report at the time the work was being conducted were the ones considered most relevant to the scope of work. There have also been new revisions to the reports cited that have not been reviewed or considered because they were issued after this work was conducted. There are some other BWRVIP reports on inspection that may be of interest that include BWRVIP-42-A (2005) on LPCI, BWRVIP-49-A (2002) on instrumentation penetrations, BWRVIP-76 (1999) on core shrouds, BWRVIP-139-A (2009) on steam dryer, BWRVIP-180 (2007) on access hole covers, and BWRVIP-183 (2007) on top guide grid beam. The BWRVIP has continued to work and has created a total of 235 reports addressing BWR vessels and internals. These reports are listed on the EPRI website www.epri.com.

2.2 Degradation and Inspections

Topics for this section include utility experiences with SCC detection, the BWR Vessel and Internals Project (BWRVIP), and the ASME Code Section XI (2007) for the inspection of reactor internals. Vendor services for ISI of reactor internals are also covered. The owners of BWRs started the BWRVIP to coordinate efforts in managing the degradation (including SCC) of reactor vessels and internals.

The BWRVIP is a program organized into four tasks: inspection, assessment, mitigation, and repair. This program has produced separate inspection guidelines for safety-related reactor internal components, including core shroud (BWRVIP-03), top guide (BWRVIP-26), core plate (BWRVIP-25), jet pump assembly (BWRVIP-41), core spray internals (BWRVIP-18), RPV attachments (BWRVIP-48), shroud support (BWRVIP-38), liquid standby control (BWRVIP-27), and lower plenum components (BWRVIP-47). The BWRVIP is intended to serve as the focal point of the regulatory interface for industry on BWR vessel and internals operability issues.

The inspection task of the BWRVIP is evaluating NDE inspection techniques including visual testing (VT) and ultrasonic and eddy current methods. For accessible components, many field SCCs have been detected and length-sized by visual NDE methods, and an assessment of length sizing uncertainty was conducted using industry mockups. These mockups have been constructed for assessing and developing ultrasonic and eddy current testing of reactor internals.

Internal components of BWRs that have experienced degradation are listed in *Vessel Internals Inspection Summary* (BWRVIP 1997). Table 2-1 shows the NDE techniques used in the in-service inspection of BWR internals. Inspections have detected SCC in the stainless steel shrouds, and cracking has been detected in other internal components that are composed of a variety of stainless steel alloys. The cracking occurs in the heat-affected zone of welds where grain boundary sensitization, residual stresses, and the corrosion chemistry produced by the radiation environment interact to cause SCC.

Table 2-1   Techniques Used in In-service Inspection of BWR Internals as Cited in Available Literature

Component	Material	NDE Technique	Sources
Core shroud assembly	Wrought austenitic stainless steel	VT, UT	[1], [2], [3], [4]
Top guide assembly	Wrought austenitic stainless steel	VT	[5]
Core plate assembly	Wrought austenitic stainless steel	VT, UT	[6]
Jet pump assembly	Wrought austenitic stainless steel	VT, UT	[7], [8], [9], [10], [11]
Core spray internals	Wrought austenitic stainless steel	VT	[12]
RPV attachments	Alloy 182	VT	[13]
Shroud support assembly	Alloy 600/182	VT, UT	[14], [15]
Liquid standby control	Alloy 600/182	VT	[16]
Lower plenum components	Wrought austenitic stainless steel, Alloy 600, cast austenitic stainless steel	VT	[17]

[1] BWRVIP-03 (1995)   [10] NRC (1980)
[2] Baker and Van Hoomissen (1994)   [11] Baker et al. (1994)
[3] Bertz et al. (1994)   [12] BWRVIP-18 (1996)
[4] Hayden (1998)   [13] BWRVIP-48 (1998)
[5] BWRVIP-26 (1996)   [14] BWRVIP-38 (1997)
[6] BWRVIP-25 (1996)   [15] McDonald (1994)
[7] Hacker et al. (1998)   [16] BWRVIP-27 (1997)
[8] BWRVIP-41 (1997)   [17] BWRVIP-47 (1997)
[9] GE Nuclear Energy (1996)

Examples of stress corrosion cracks that were obtained by laboratory and field-induced SCC in piping welds can be seen in Good and Van Fleet (1986). The illustrations provided in Good and Van Fleet (1986) show the openness and features that can form in SCC. The cracks show branching and grain encirclement. These crack properties influence the reliability of UT-ISI. The branching and grain encirclement along the crack face scatter ultrasound. This scattering of ultrasound from such a rough crack face affects the ultrasonic images, as described in Sections 5 and 6.

Figure 2-5 depicts representative weld microstructure of a core shroud weld (permission to use this copyrighted material is granted by the Electric Power Research Institute). The microstructure of the weld metal significantly impedes the propagation of ultrasound. The results of the parametric study, given in Section 6, show the significance of weld microstructure in the detection and sizing of SCC in the heat-affected zone of welds in reactor internals.

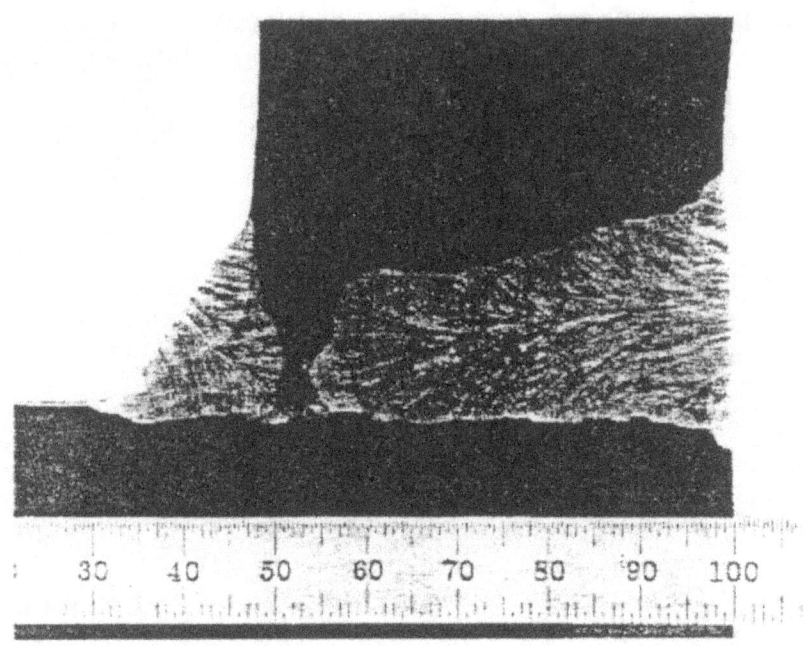

Figure 2-5   Typical Microstructure Content for a Boiling Water Reactor Core Shroud Weld. Source: BWRVIP-03 (1995), © 1995 Electric Power Research Institute, Inc., reproduced with permission.

Intergranular stress corrosion cracking in austenitic piping was a major issue for BWRs in the 1980s. IGSCC had a significant impact relative to outages, occupational exposure, and capacity factor losses. In addition, primary piping systems were designed and are inspected in accordance with the ASME Code, which is incorporated by reference into the NRC regulations. Thus, measures were quickly implemented to address IGSCC in primary piping.

The Electric Power Research Institute (EPRI) and the BWROG recognized the potential susceptibility of reactor internals to IGSCC. IGSCC of reactor internals became a reality in the 1993–1994 time frame when examinations at certain nuclear power plants detected cracks in core shrouds. However, unlike primary piping, most BWR internal components are not designed or inspected in accordance with the ASME Code (the ASME Code was developed to establish standards of safety related only to pressure boundary integrity). The BWRVIP, an association of utilities, was formed by BWR utility executives in mid-1994 to proactively address, on a generic basis, issues related to reactor vessel and internals material. The BWRVIP established agreements with the NRC regarding commitments to the program because vessel internals were not adequately addressed by the ASME Code.

As the BWRVIP has matured, other objectives have been established, such as to identify or develop generic, cost-effective strategies; serve as a focal point for the regulatory interface with the industry; and share information among utilities. The development of generic solutions by the BWRVIP to address the degradation of vessel internals in conjunction with the regulatory commitments meant that the NRC did not have to expend considerable rulemaking resources to deal with internals degradation.

The BWRVIP addresses issues through the issuance of guideline reports. There is an inspection and evaluation guideline document for each internal component addressed by BWRVIP. These reports consolidate various procedures developed by the industry as well as information from NRC reviews and safety evaluations (SEs). The purpose of the guidelines is to ensure that a unified industry approach, which has been reviewed and accepted by the regulator, will be implemented for evaluating and maintaining the integrity of vessels and internals.

To date, over 150 guideline reports have been developed. A safety assessment was performed by the BWRVIP to identify which components needed to be addressed and to establish examination schedules. The safety assessment identified the components that are necessary for safe operation and shutdown of the reactor to maintain a coolable geometry, rod insertion times, and reactivity control, and to ensure instrumentation availability. It describes which locations are likely to experience degradation through IGSCC or other mechanisms and the safety consequences of failure for each location. It was determined that the following components needed to be addressed: reactor pressure vessel, core shroud, shroud support, core spray internals, jet pump assembly, top guide, core plate, lower plenum components, vessel inner diameter brackets, standby liquid control, low pressure coolant injection couplings, and instrument penetrations. The safety assessment also addressed issues such as re-inspection schedules, flaw evaluation, and crack growth.

It was determined that an inspection focus group was needed to address best inspection practices and the development of new inspection methods, oversee inspection demonstrations, and deal with nondestructive examination uncertainties. Two other important groups were formed: the repair focus group to support the development of repair techniques and to define technical requirements for performing acceptable repairs; and the mitigation committee to establish and demonstrate methods for the mitigation of IGSCC.

To ensure compliance with NRC requirements, the inspection and guidelines reports addressing safety-related components are submitted to the NRC for review and approval. Utility BWRVIP programs must conform to the quality assurance requirements in Title 10 of the Code of Federal Regulations, Part 50, Appendix B, "Quality Assurance Criteria for Nuclear Power Plants and

Fuel Reprocessing Plants." Upon completion of the review of each report, the NRC issues a safety evaluation (SE). A SE provides the NRC's conclusions relative to the BWRVIP assessment, as well as the limitations, if any, under which the approval is granted. Licensees participating in the BWRVIP and relying on a given report must commit to the program defined therein and complete the action items described in the SE. The NRC has determined that a commitment to a BWRVIP report and the SE provides reasonable assurance that the applicant will adequately manage degradation so that the intended functions of the component will be maintained consistent with the current licensing basis.

The BWRVIP program requires utilities to provide inspection results to the NRC. Any required repairs resulting from inspections must be qualified and implemented in accordance with BWRVIP guidelines. Utilities are required under this program to perform self-assessments to ensure compliance with program requirements. Independent assessments of individual nuclear power plant BWRVIP programs are conducted by the Institute of Nuclear Power Operations (INPO). INPO is a not-for-profit organization established by the nuclear power industry in December 1979 to promote the highest levels of safety and reliability in the operation of nuclear electric generating plants. This is achieved by establishing performance objectives, criteria, and guidelines for the nuclear power industry; conducting regular detailed evaluations of nuclear power plants; and providing assistance to help nuclear power plants continually improve their performance. Significant INPO findings are incorporated into the BWRVIP guidelines to address the issue(s) raised.

3 IN-SERVICE INSPECTION OF REACTOR INTERNALS USING ULTRASONIC TESTING

This section provides a description of the ultrasonic techniques that are applied for in-service inspection of BWR reactor internals. The ultrasonic modalities are discussed with their strengths and weaknesses. The use of 45° shear wave probes is described, and an explanation of sizing using crack tip diffraction is given. High-angle (60°–70° from surface normal) compression (longitudinal) waves are useful when inspecting through weld metal. Creeping wave probes are described and some signal interpretation is given. The time-of-flight diffraction (TOFD) technique propagates ultrasound through the weld metal in a pitch-catch arrangement and can provide accurate sizing. Phased-array techniques are being applied with increasing frequency in nuclear applications.

3.1 Ultrasonic Modalities

The most commonly used ultrasonic in-service inspection modalities are listed in Table 3-1 with their respective strengths and weaknesses. Quantitative response data are provided in Section 6 for some of the modalities. For TOFD, an extended explanation is provided in this section. The results of a blind test of a phased array procedure are given in Section 7 of this report.

Table 3-1   Relative Strengths and Weaknesses of Ultrasonic Modalities for Inspection of Austenitic Welds of Reactor Internals

Modality	Characteristic Strengths	Characteristic Weaknesses
45° shear wave	Near-side detection and sizing	Insensitive for far side cracks
60° to 70° compression wave	Cracks connected to scanning surface and diffraction sources on far side	Insensitive to corner trap for far-side cracks
60° to 70° shear wave	Corner trap for far side cracks	Mode converted signals produce image artifacts
Creeping wave	Sensitive to surface-connected discontinuities	Multiple modalities require extra interpretation
Time-of-flight diffraction	Accurate sizing	Transmission through weld metal is required
Phased array	Multipath acoustic illumination	Relatively low sensitivity or dynamic range

Figure 3-1 shows some typical commercial transducers used for inspecting austenitic materials. The transducer on the left is a near-surface, dual-element, 4.0-MHz, 45° refracted longitudinal wave probe with a 25-mm focal depth. Second from left is a near-surface, dual-element, 4.0-MHz, 60° longitudinal wave probe with a 20-mm focal depth. Third from the left is a far-surface, dual-element, 2.0-MHz creeping wave probe with a 50-mm focal depth. Fourth is a far-surface, dual-element, 2.0-MHz, 60° refracted longitudinal wave probe with a 75-mm focal depth. On the right is a far-surface, single-element, 2.25-MHz, 45° shear wave probe.

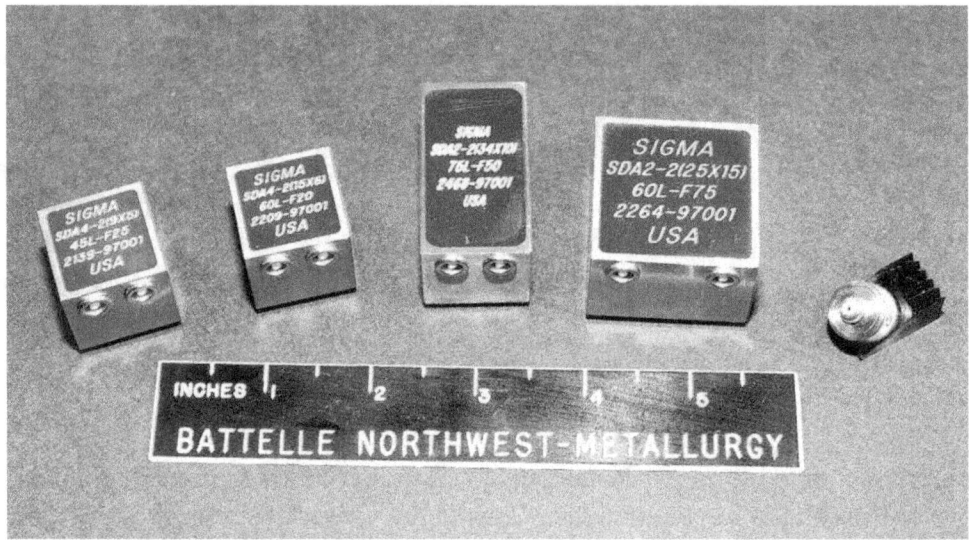

Figure 3-1 Commercial Transducers Typically Used for ISI of Austenitic Materials

3.2 45° Shear Waves

Ultrasonic examination for heat-affected zone cracking using vertically polarized 45° shear waves is represented in Figure 3-2, which shows the annulus side of the core shroud on the right and the fuel side on the left. Most inspections of the core shroud will be made from the annulus side. Cracks form on either side of the weld and can originate from the annulus or from the fuel side. The weld metal is shown as a number of individual weld passes with radially oriented grains. The weld grains, coarse compared to the parent material, are a significant scattering source of ultrasonic energy. The scattering depends strongly on grain size and orientation and on the wavelength of the ultrasound. The propagation of 2.25-MHz vertically polarized shear waves is impeded significantly by the weld grains, making far-side inspection unreliable compared to the inspection reliability for cracks located on the near side.

Sizing with a 45° shear wave probe is accomplished using ultrasonic energy diffracted from the crack tip, as shown in Figure 3-3. As illustrated, the transducer is moving toward the crack in the three sketches. The top sketch shows the center ray of the transducer striking the corner of the crack with the back wall of the test piece. For 45° shear waves, this corner trap signal will be bright, generally more than 20 dB brighter than other signals from the crack. The middle sketch shows the center ray of the transducer striking the face of the rough crack. Sound will be diffracted from the face of the rough crack and recorded by the transducer. The bottom sketch shows the center ray of the transducer striking the tip of the crack. Sound will be diffracted from this tip, and it can be used to provide an estimate of crack size.

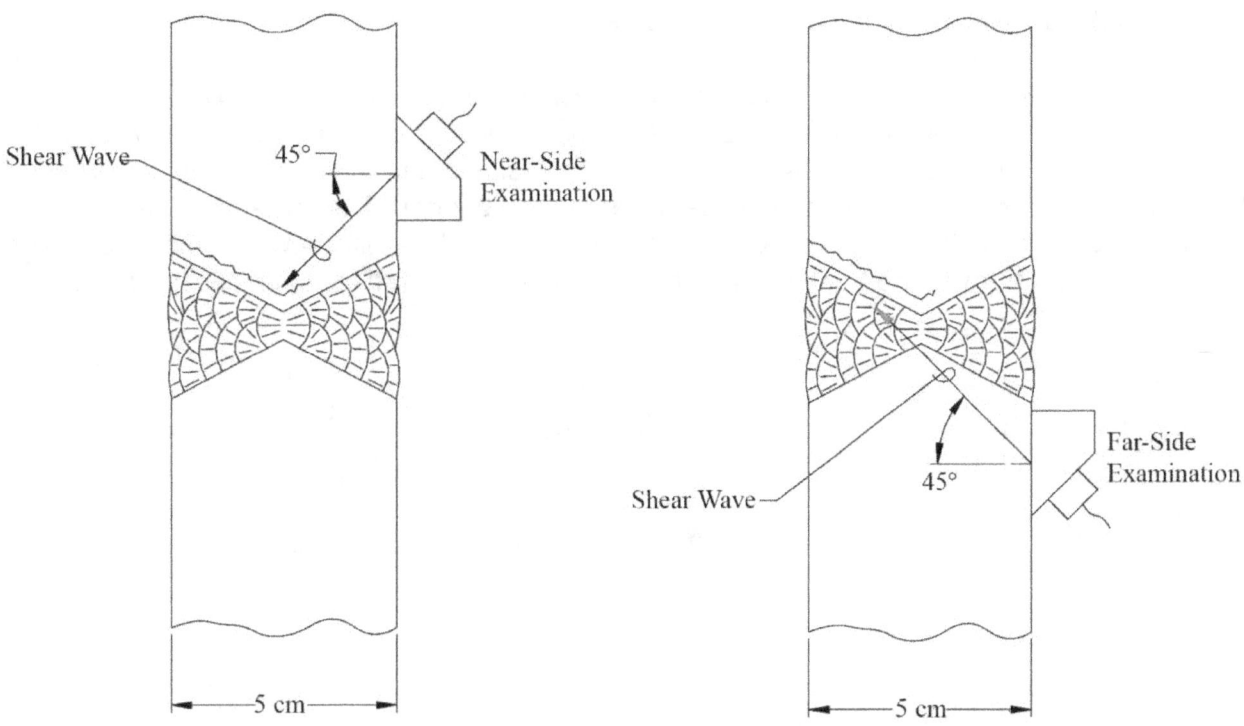

Figure 3-2 **Near-Side Access (left) and Far-Side Access (right) Ultrasonic Examination Using Vertically Polarized 45° Shear Waves for Cracks That Are Opposite-Surface Connected (from the transducer)**

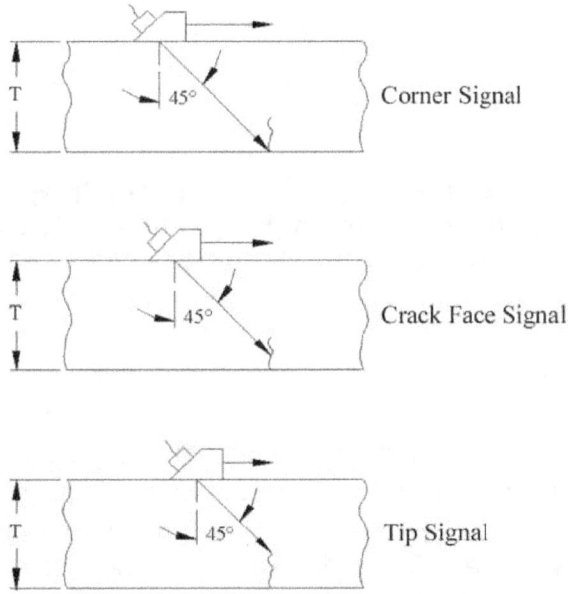

Figure 3-3 **Three Ultrasonic Signals (Corner Trap, Crack Face, and Crack Tip) from Opposite-Surface Connected Rough Cracks**

3.3　High-Angle Longitudinal Waves

High-angle (60° to 70° from surface normal) compression (longitudinal) wave examination is shown in Figure 3-4. The inspection of a K weld profile, typical for a ring-to-cylinder weld in the core shroud, is shown. The crack is located on the ring side of the weld. Inspections are shown for near-side access on the top and far-side access at the bottom. The crack is connected to the same surface that is used for conducting the inspection. Data will show that the crack faces return acoustic energy to the transducer. Little or no corner trap reflection is measured in this case.

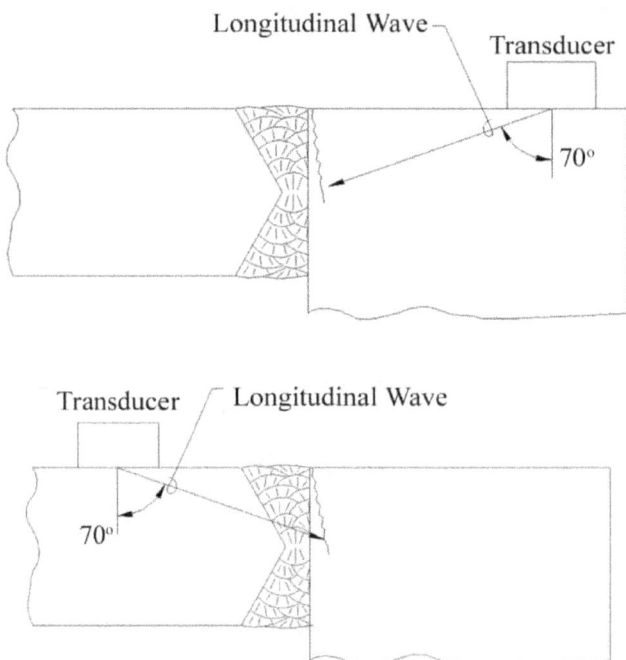

Figure 3-4　Near-Side (Base Metal Access) Shown in Top Image and Far-Side (Weld Metal Access) Shown in Bottom Image Ultrasonic Examination Using High-Angle Compression Waves for Scan-Surface Connected Cracks

3.4　Creeping Waves

The creeping wave technique is a variation of the high-angle longitudinal (L) wave that introduces multiple ultrasonic modalities into the component. Figure 3-5 shows the multiple modalities transmitted by a creeping wave probe. The transducer and angled elements that introduce 75° longitudinal waves and 33° shear waves into the test material are depicted. For cracks that are opposite-surface connected, three insonifications must be considered. As the creeping wave probe is translated toward the crack, the first insonification will be from the 75° main longitudinal wave. As the transducer approaches the crack, the secondary 75° longitudinal waves will insonify it. Finally, the 33° shear waves insonify the crack, and a corner trap reflection is recorded. Thus, a complex but unique response is produced. This techniques requires good access to allow all three modalities to insonify the entire inspection zone.

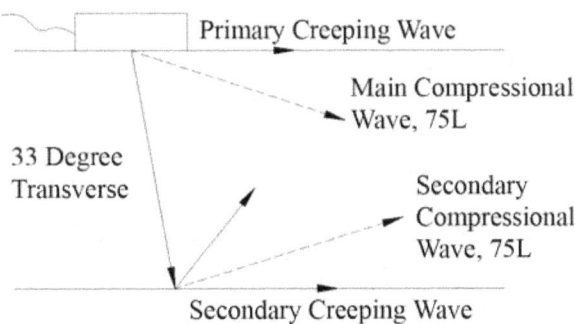

Figure 3-5 Creeping Wave Technique

3.5   Time-of-Flight Diffraction Technique

In the TOFD technique, a pair of transducers is placed astride a weld, and flaws throughout the volume of the weld and heat-affected zone are insonified as the probe pair is traversed along the weld axis. Flaws of all orientations and types are detected and displayed in real time (except where no tip signal is present). The limiting factors are access to both sides of the weld and propagation of ultrasound through weld metal.

The TOFD technique differs from that of the conventional pulse-echo ultrasonic inspection in two major ways:

- TOFD has an improved ability to reliably locate and size flaws in unfavorable orientations.

- TOFD measurements of depth and through-wall extent rely only on the ability to detect and measure signal time of arrival.

Typical pulse-echo techniques are sensitive to parameters such as probe position, beam spread, incident and skew angles, and flaw roughness and orientation. Because TOFD is a forward-scattering technique, it can be much less sensitive to these variables. For these reasons, TOFD can be used for accurately locating and sizing flaws and for monitoring flaw growth. Work performed by AEA Technology shows the sizing accuracy that TOFD provides (Lilley 1994). However, interpretation can be challenging since only tip signals are detected. For example, in cases where two small flaws are stacked above one another, will they be correctly interpreted as two small flaws or one large flaw?

The approach is based on getting two responses: (1) obstruction of the lateral wave or back wall echo for surface-connected cracks and (2) tip-forward scattering. Figure 3-6 shows the transmitting and receiving transducer pair used in TOFD. The technique is shown producing four responses in the receiving transducer. The first to arrive is the lateral wave. Such a lateral wave is always present in the inspection data when no cracks are connected to the scanning surface. The cracks tips (upper and lower) diffract acoustic energy sensed by the receiver.

These diffraction signals are used to detect, locate, and size cracks in reactor internals. Finally, a back-wall echo should be available for the case in which no cracks are connected to the opposite surface.

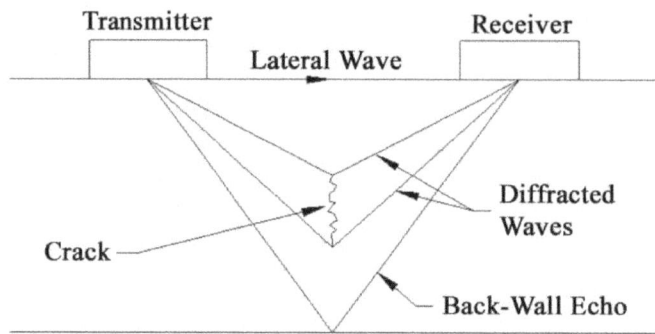

Figure 3-6 Time-of-Flight Diffraction Technique

All responses must propagate through the weld metal. Sufficient scanning surface must be available for the motion of both the transmitter and the receiver. For reactor internals, fillet welds and other obstructions often limit the inspection.

3.6   Phased Arrays

The phased array technique refers to a process in which ultrasonic data are collected by continuous and incremental variation of the ultrasonic beam angle, typically in azimuthal direction, while scanning the object under investigation. This technique potentially offers advantages over conventional techniques employing search units with fixed beam angles because a considerably greater amount of information can be obtained about the shape and orientation of reflectors in the test material.

Each phased array search unit consists of an array of individual ultrasonic transmitting and receiving elements. These elements are activated separately in a defined, pre-selected time delay pattern to obtain the desired beam angle originating in a plane perpendicular to the separation of the individual crystal elements. A linear time delay between the stimulation of the different elements results in an inclined sound field, while a variation of the linear time delay distribution results in a variation of the angle of refraction. Using a curved time delay pattern similar to the shape of an optical lens, a focused sound field at a certain distance from the elements is generated. Changing the radius of the optical lens will change the focus of the sound field; therefore, a controlled variation of the curved time delay pattern allows a variance of the focal length of the sound field within the near-field distance for the transducer site. The combined time delay pattern enables the phased array system to electronically steer the sound field. Different search unit designs allow for electronic variation of either the incidence angle or the lateral angle. The same time-delay pattern utilized for transmitting is used in the receiving mode.

The phased array system can operate using either shear or compressing (longitudinal) waves. A phased-array system is capable of employing a number of scanning techniques and is controlled remotely using a variety of manipulator systems.

3.7 Deployment

The inspection problems inherent in ISI of BWR internals arise from many factors, including component geometry, effectiveness of inspection technique, inspection hardware (scanner mechanisms), material type and thickness, surface finish, and limitations with regard to space and accessibility. BWR plants have experienced significant inspection challenges and identified cracking in access hole covers, steam dryers, jet pump beams, instrument line supports, and the core shrouds. Because owners are responsible for addressing their specific inspection issues, they prepare and implement internals inspection programs. To select the most effective inspection methodology, a variety of factors need to be considered, including materials, susceptibility to cracking, applied stresses, radiation fields, accessibility, and available inspection techniques. Planning for internals inspection requires careful and detailed technical evaluation.

Manipulation (mechanical) systems have been developed for field inspections of internals components. The WesDyne Inc. system for automated inspection of the vertical and horizontal welds in a BWR core shroud is described in Davis and Huntington (1998). Figure 3-7 shows a core shroud examination using the Siemens Power Corporation lower manipulator (Fisher and Tagliamonte 1994; permission to use this copyrighted material is granted by the Electric Power Research Institute). The inspection is conducted from the outside surface of the core shroud. The scanning device uses the vessel wall for support. Horizontal telescoping of the search unit is provided to increase coverage and compensate for mechanical interference from the jet pumps.

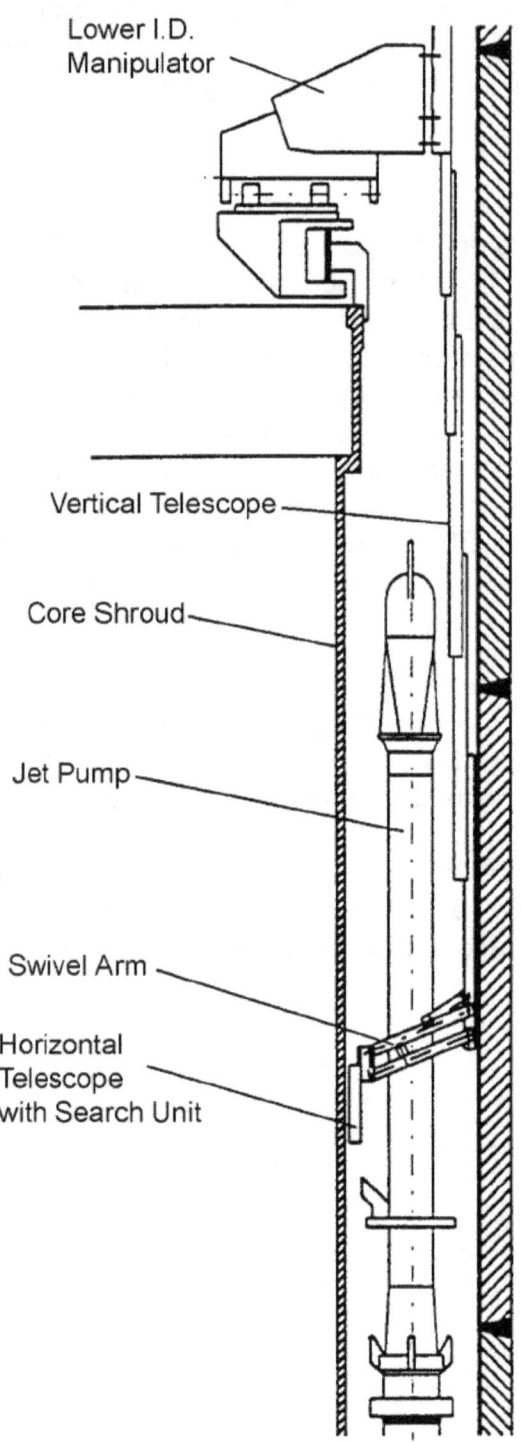

Figure 3-7   Core Shroud Examination Using the Siemens Power Corporation Lower Manipulator. Source: Fisher and Tagliamonte (1994), Electric Power Research Institute, Inc.; reproduced with permission granted by EPRI.

4   RELIABILITY METHODS

The ability of nondestructive testing to reliably detect and accurately size stress corrosion cracks in stainless steel piping has been studied in research programs for about 25 years. Measurement of the performance of standard and advanced NDE practice has been made in PNNL round robins and in the international Programme for the Inspection of Steel Components (PISC III). This chapter summarizes the results from these studies.

Figure 4-1 shows probability of detection (POD) versus through-wall extent of flaws in various studies, over the past two decades, using logistic regression. The lowest performance level is shown for the Piping Inspection Round Robin (PIRR) (Heasler and Doctor 1996) study that took place in 1981. This study is the benchmark of performance in the early 1980s, meeting the 1970s ASME Code Section XI prescriptive inspection requirements. It was just after the PIRR that the issue of inspection reliability became high priority for austenitic welds. The mini-round robin (MRR) (Heasler et al. 1990) results in 1985, following industry-developed training and initial performance demonstration testing, showed that smaller flaws could be more reliably detected with a moderate increase in performance for intermediate flaws. The PISC-AST (Lemaitre 1994) results of 1990 show a significant improvement over the PIRR and, except for very small flaws, show an improvement over the MRR results. There are two results for the Performance Demonstration Initiative (PDI) data (Doctor and Becker 2002), which reflect the performance demonstration testing that is in the ASME Code Section XI, Appendix VIII. The PDI data are plotted for all inspectors who successfully passed the performance demonstration requirements and a second curve for all inspectors who attempted the performance demonstration test (passed and failed). It can be seen that both of these PDI-based curves show improvements over the preceding three round robin studies.

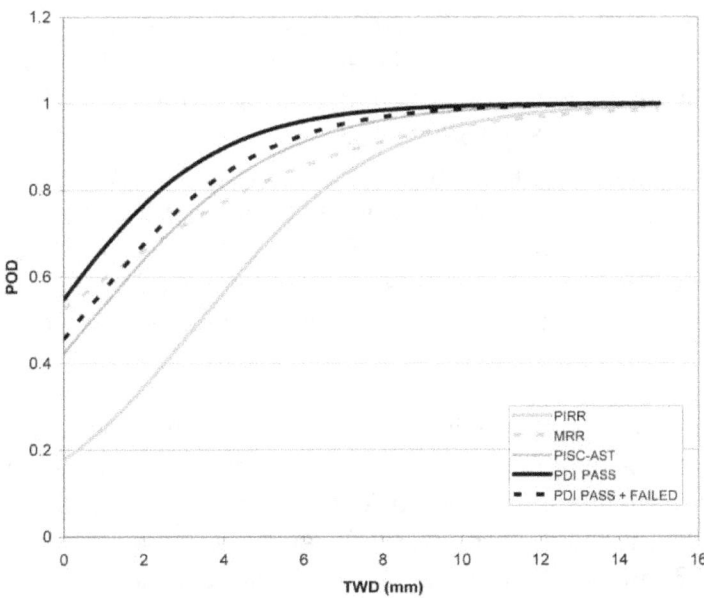

Figure 4-1   Changes in POD versus Through-Wall Dimension of Flaws for Piping Weld Inspection in the Past Two Decades (Doctor and Becker 2002)

4.1 Blind Tests

Performance in blind tests can provide an estimate of two relevant crack detection statistics of an inspection procedure: POD and false call probability (FCP). The POD is the probability that a crack in a specific length of weld material will be detected by the inspection. The FCP is the probability that an inspection will characterize a similar unit of blank material (defect-free) as cracked. The tests typically consist of a fixed number of blank and cracked grading units that are presented in random order.

The size of the blind test is determined by its primary objective—to identify procedures/equipment/personnel that can correctly characterize fixed units of a welded component. It can be shown that a test with 10 cracked grading units and 20 blank ones is a powerful test for screening inspection capability (Heasler et al. 1986; Heasler et al. 1990; Heasler and Doctor 1996). If the pass/fail criterion is detecting 8 of the 10 cracks and having not more than 3 false calls, then an inspection with an assumed POD = 90% and an FCP = 10% has about a 90% chance of passing the test. An inspection with an assumed POD = 50% and a FCP of 30% has only about a 1% probability of passing the test. It is important that the cracked and blank grading units have a representative amount of part noise and interference signals. Part noise is back-scattered ultrasonic signal from the metal's microstructure and surface geometry (both inside and outside the part). Interference signals are an important source of systematic noise.

In the field, the unit of material that must be characterized is a weld. Clearly, it is impractical for a blind test to use this basic unit. Furthermore, weld units of different sizes cannot be used easily in this analysis because comparable statistics can be calculated from only blank and cracked weld units of the same size. To avoid confusion and allow for simpler test results, it is best to use the same unit of material, called the grading unit, to estimate FCP and POD.

The comparison of POD and FCP performance of inspection can be plotted on what is called an operating characteristics (OC) diagram. These OC diagrams frequently are employed to evaluate all sorts of diagnostic systems, particularly in the medical field (Swets and Pickett 1982; Swets 1983). Points in the OC diagram can represent the performance of individual procedure/equipment/personnel. Figure 4-2 shows an example of a typical OC diagram. The performance of each procedure/equipment/personnel should fall above the diagonal line in the plot. The diagonal line represents the performance that could be achieved with random guessing and therefore represents the worst performance that a procedure might achieve. The best performance in the diagram is represented by the point (FCP = 0, POD = 1), which indicates that a procedure/equipment/personnel discriminates perfectly between cracked and blank grading units.

The OC diagram shows that it is not sufficient to examine POD alone. For example, the performance of Inspection A in Figure 4-2 is obviously better than Inspection B, if we consider only POD. But the performance of Procedure A lies on the diagonal, which shows that Inspection A's high POD (and high false call rate) was achieved by classifying blank material as flawed. Inspection B, on the other hand, has demonstrated the ability to more effectively distinguish between cracked and blank material.

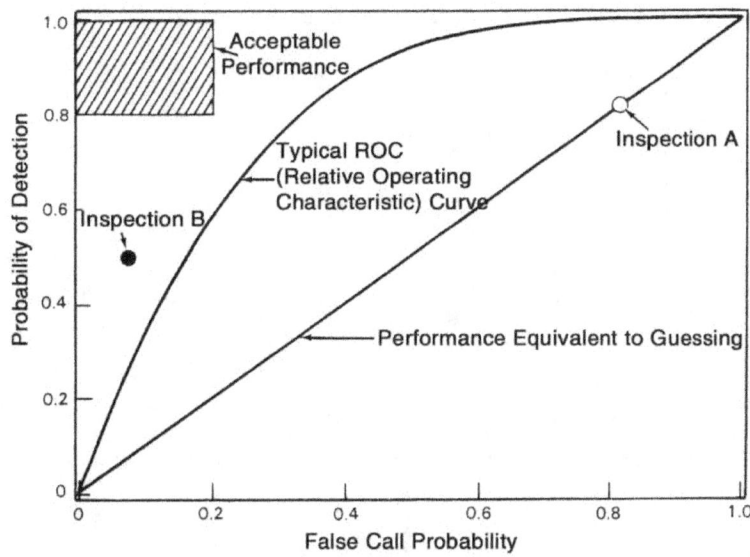

Figure 4-2 Operating Characteristics Diagram (Heasler et al. 1986)

In the above discussion, we have assumed that fixed decision criteria were used. In fact, an inspector can change the decision criteria, and the performance can vary along what is called the relative operating characteristic (ROC) curve shown in Figure 4-2.

4.2   Round Robin Tests

The PISC III conducted studies of ultrasonic inspection of stainless steel piping welds. The objective was to identify high-performance inspection techniques and procedures and to inform codes and standards organizations of the results obtained (Lemaitre et al. 1996). The study investigated the influence of geometrical conditions, such as counterbore, on detection performance. Different metallurgical conditions were investigated as well as flaw type. An ASME Code Section XI, Appendix VIII type of performance demonstration test was simulated.

Figure 4-3 shows the detection performance for procedures in the inspection of stainless steel welds in PISC III. Half of the teams detected more the 75% of the flaws. All of the flaws above 7 mm in depth had a flaw detection frequency (FDF) greater than or equal to 0.9.

The best techniques used both compression and shear waves. Shear wave techniques were more sensitive than compression wave techniques and enabled detection of more difficult flaws. Shear wave techniques generated more indications from geometry. Operator skill in combining information from shear and compression wave probes was the key to achieving good detection and low false calls. A number of teams had a high number of false calls that exceeded the acceptance criteria.

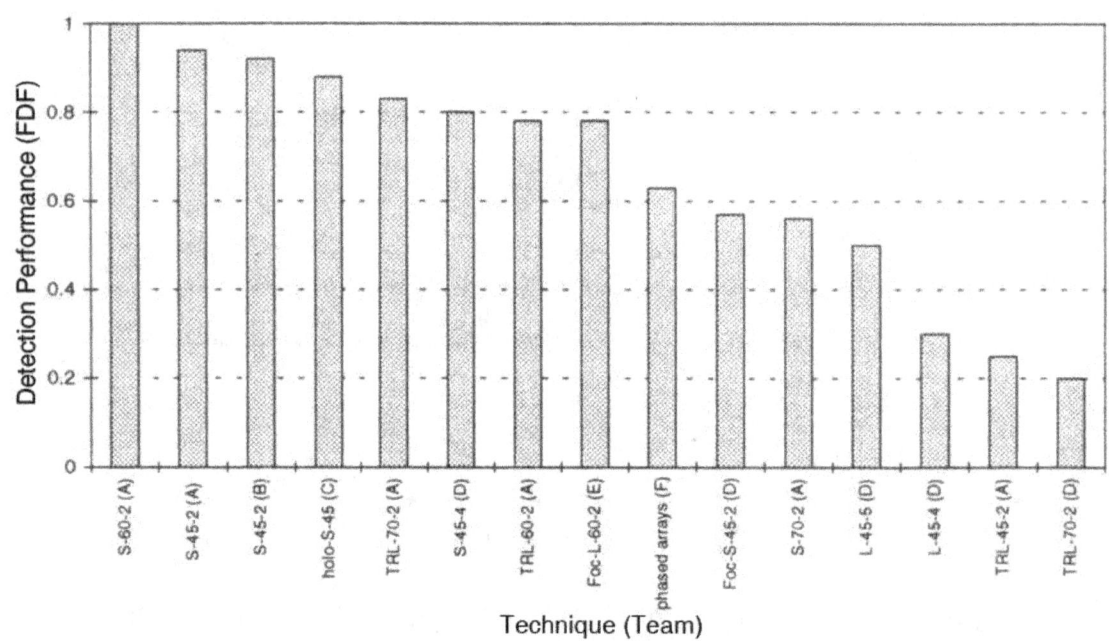

Figure 4-3   Flaw Detection Frequency for Procedures in PISC III, Action 4 (Lemaitre et al. 1996)

Figure 4-4 provides the OC diagram for techniques in the inspection of stainless steel welds in PISC III. Large scatter in results by teams that used similar procedures (techniques) show that they should be applied by skilled and experienced operators. Training is needed on realistic specimens. Access to both sides of the weld is needed. Scanning in two different directions, perpendicular to the weld, improved detection performance.

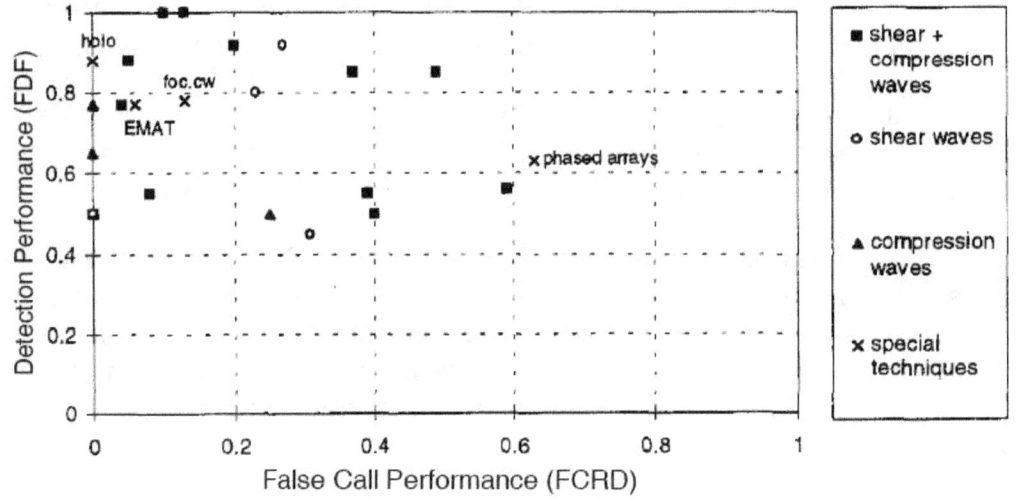

Figure 4-4   Operating Characteristic Diagram for Techniques in PISC III, Action 4 (Lemaitre et al. 1996)

The five teams that did well on all categories used an automated procedure and worked at the noise level of the test specimens. Good depth-sizing techniques used crack-tip diffraction. Amplitude drop methods did not work reliably.

4.3 Parametric Studies

Previous work conducted on parametric performance of ultrasonic testing of piping welds is summarized in this section. In parametric studies conducted by PNNL, the estimates of responses and their variances are made in stainless steel piping (Becker et al. 1981). The data from the parametric study are included in this section to show that 6 dB of response variance can be expected in UT-ISI.

Estimates for various sources of response variance are listed in Table 4-1 along with estimates made by Silk (1978) and Forli (1979a, b). The standard deviations for the overall inspection process are similar, although they are derived from different sources. The data reported by Silk resulted from a compilation of approximately 16 different experiments on fatigue cracks in aluminum and steel. The data by Forli are the result of a round robin test on 70 m of mild steel butt welds, 10 to 26 mm thick. The study reported by Forli employed 12 inspection teams and 60 cracks, which were predominantly fabrication flaws reflecting lack of root penetration and lack of fusion.

Table 4-1 Estimate of Standard Deviation of Inspection Variables (Becker et al. 1981)

	Becker et al. (1981), ±dB	Silk (1978), ±dB	Forli (1979a,b), ±dB
σ_3 Within inspection	2	2	4.3
σ_2 Between inspections	2	-	2.7
Coupling	-	2	-
σ_1 Crack-to-crack	5.4	6.1	3.3
Crack orientation	3.6	3.5	-
Crack roughness	1.4	3.0	-
Crack transparency	3.7	4.0	-
σ_T Total inspection	6.1	6.7	6.1

The within-inspection error is probably the most well-documented of all the variables. The PNNL data are based on approximately 500 measurements under laboratory conditions, while Forli's 586 measurements represent field conditions. Forli's data were collected at 10%, 20%, and 50% of a calibration response, which may account for the higher within-inspection variability. PNNL's estimate of 2 dB, for between-inspector standard deviation, is an engineering judgment based on the differences expected due to search unit and instrument selection. Silk did not estimate between-inspector errors; however, the variable is accounted for in the total estimate. Silk also provides an estimate of ±2 dB for coupling errors that are not evaluated separately by the other two studies.

The total error due to the flaw character is made up of the contributions due to defect orientation, roughness, and transparency. Only the total flaw contribution, 3.3 dB, is available from the data by Forli. Because the Forli data are based on fabrication flaws, the major contributors are most likely orientation and roughness, with transparency having little influence. The estimates from the PNNL study are slightly less conservative than those presented by Silk, particularly in regard to flaw roughness effects. This difference exists partly because a portion of Silk's data was measured at 5 MHz instead of the 2.25 MHz used in the PNNL study. Increased frequency increases the scattering effects due to roughness.

Figure 4-5 shows the effect of a 75% compressive stress on the measured signal response. Although a 75% yield compressive stress may appear very conservative, it should be noted that the data are for dry fatigue cracks. Liquid-filled cracks, SCCs, and tight thermal fatigue cracks may produce larger losses in signal amplitude with no external stress applied. The experimental data are shown in Figure 4-5, along with the response curve for ideal reflectors of aspect ratio 0.2 and the fitted curve. The mean of the measured flaws is 7 dB below the ideal response curve, with a standard deviation of 4 dB. The 2σ error bars for this measurement also are shown in Figure 4-5.

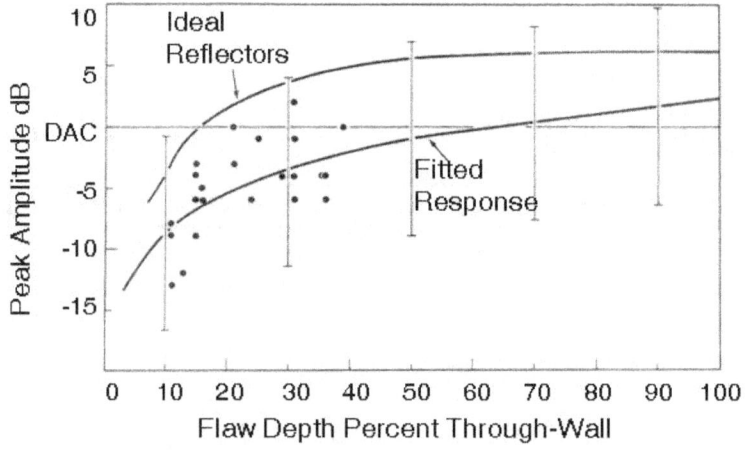

Figure 4-5 Response versus Crack Tightness (Becker et al. 1981)

5 REACTOR INTERNALS CORE SHROUD MOCKUP DESIGN

In this section, the reactor internals mockup is presented with the data used to design it. A description of BWR reactor internals and the core shroud weld configuration was provided in Section 2. Sections 3 and 4 provided an explanation of the ultrasonic modalities applied to reactor internals and the measurement of reliability. Here, we describe the development of some terms for an analysis of possible mockup conditions, our tests of the mockup design to support both a parametric study of NDE reliability and a blind test of vendor performance, and the assemblies that were built. Illustrations are presented that describe the use of ultrasonic testing on core shroud conditions and how these conditions are represented in the mockup.

The focus of this study is the quantification of inspection reliability of the vertical and horizontal core shroud welds. Because BWR internals have different designs, the access restrictions and therefore ultrasonic coverage will vary from plant to plant. The mockup described here is representative in that it provides a range of access conditions to a crack or degraded material that was found to have a significant effect on NDE reliability. This section provides an in-depth discussion of the component conditions studied, types of simulated flaws studied, the acceptance test employed, and a complete description of the mockup parameters. Quantitative performance of the ultrasonic techniques using the mockup is given in Section 6. The statistics that were acquired in a blind test of an ISI vendor's phased array system are given in Section 7.

5.1 Component Conditions

The inspection areas of interest include the heat-affected zone of the horizontal and vertical welds in the core shroud. Inspection from the outside of the core shroud would normally be preferred because the inside is accessible only through the top guide. But the inspection modality from the outside must look though the weld material, and the weld microstructure is known to make this inspection unreliable when vertically polarized shear waves are used. A normal beam inspection from the outside might detect the crack, but is complicated by the core shroud geometry. Angle beam inspections from the support ring surface would need to use steep inspection angles (30° or less) and inspect through more than 15 cm of stainless steel.

Access conditions may limit the completeness of an outside-only inspection. The space between the core shroud and the vessel wall is not large. The distance between the two is typically 45 cm. The region between welds H4 and H9 contains the jet pumps. The geometry of the support rings can lead to insonification difficulty. The presence of weld crowns can prevent proper scan patterns.

The core shroud and support assembly is made from two types of base metal—wrought austenitic stainless steel and Alloy 600 (Inconel). The wrought austenitic stainless steel portion is above weld H7 (see Figure 2-2, where the weld numbers are identified). The nominal thickness of this portion of the core shroud is 40 mm to 50 mm. But it is also composed of 75-mm-thick support rings. The portion of the core shroud that is 40 mm to 50 mm thick is referred to as the shroud cylinder. The Alloy 600 shroud support assembly is below weld H7. The nominal thickness of the Alloy 600 is 75 mm. It is also composed of a thicker shroud support pedestal.

Table 5-1 lists the weld profiles used in BWR core shrouds. The weld profiles vary according to location in the core shroud, as described in Table 5-1. The amount of weld metal in the ultrasound path to a flaw or degradation site can influence the inspection reliability. For this reason, all of the weld profiles listed in Table 5-1 are represented in the mockup.

Table 5-1 Weld Profiles Selected for Study

Weld Profile	Typical Core Shroud Weld
Single V	Vertical welds
K	Ring to cylinder welds
Double V	Cylinder to cylinder welds
Double J	Vertical welds

The configuration of the welded assemblies for the mockup is given in Figure 5-1. The size of a welded assembly for use in the mockup is 36 cm across the weld, 25 cm along the weld, and 5 cm thick. The thickness was chosen to be representative of the BWR core shroud. The distance along the weld is enough to allow for two grading units in a blind test. The distance across the weld is enough to permit high-angle insonification of the weld's heat-affected zones. The approximate weight of such an assembly is 36 kg, light enough to facilitate manual handling by PNNL NDE staff.

Four weld profiles were selected for representation in the mockup. Figure 5-1 provides a sketch of the weld profiles and the descriptive name given to that profile, such as K weld. The weld metal in wrought-to-wrought stainless steel welded assemblies has a coarse-grained microstructure and strongly scatters ultrasound. The amount of the weld metal and the through-wall profile of the weld are important variables in the parametric study reported here.

It is interesting to note the features that are missing in the assemblies shown in the Figure 5-1. No counterbore or weld root conditions are present; those conditions generally add difficulty to the NDE reliability of the UT-ISI. The absence of these piping conditions will simplify the analysis of reliability, as described in Section 6. Typical weld crown conditions that occur in installed BWR core shrouds were included in the mockup.

Figure 5-2 shows the first five welded assemblies built to test and assess the mockup design. These assemblies—with various weld profiles, side-drilled holes, notches, and other reflectors—were used to evaluate the usefulness of the assemblies. The welds were fabricated in the vendor's controlled environment using a convenient down-hand welding position. Figure 5-3 shows the welding of one of the reactor internals mockup assemblies. The results of initial testing performed on these assemblies are reported in Section 5.3.

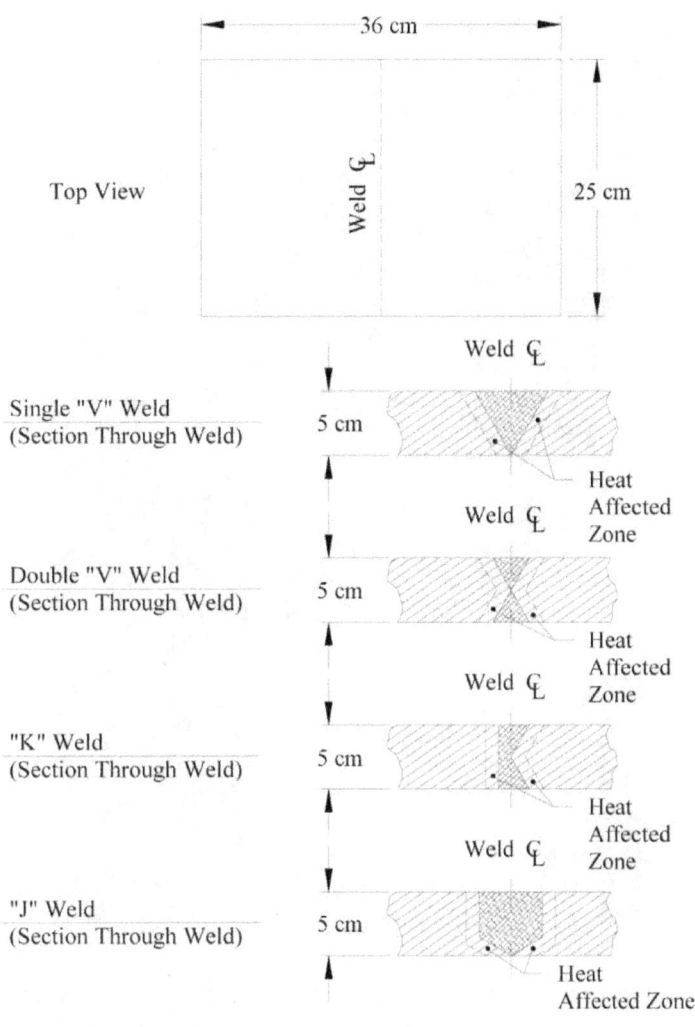

Figure 5-1 Welded Assembly Configuration for Mockup

Figure 5-2 First Welded Assemblies

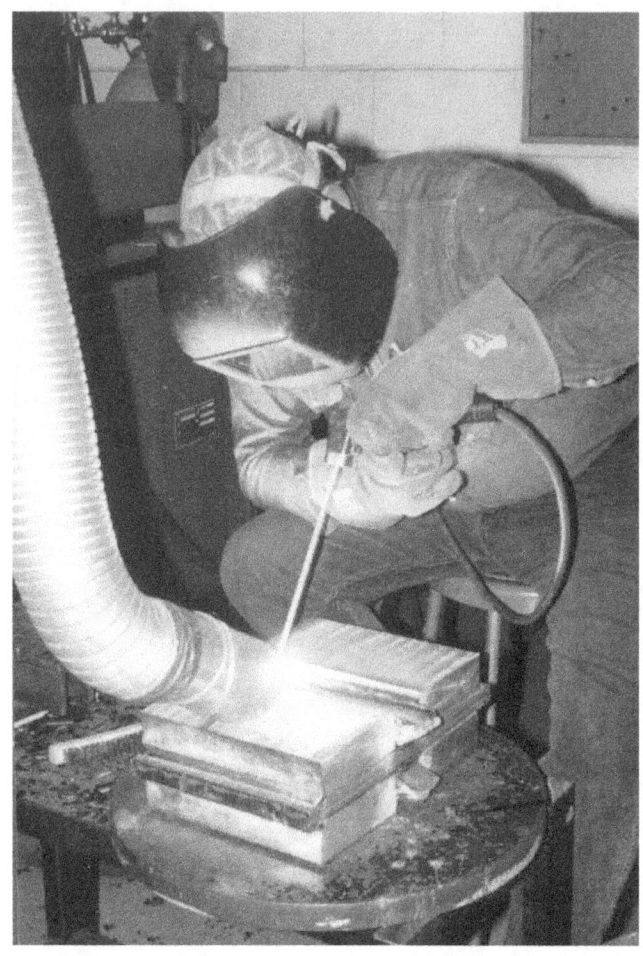

Figure 5-3   Welding of Reactor Internal Mockup Assembly (photo courtesy of FlawTech, Inc., Concord, North Carolina; reprinted with permission)

Four crack initiation sites can be expected on welds in the core shroud. Figure 5-4 shows a K-weld profile with a crack initiation from the fuel side of the top guide support ring of a BWR core shroud. In many BWR core shrouds, this weld is labeled the H3 weld. In-service inspection data show that cracks can initiate from the fuel or annulus side of the core shroud. The ring or cylinder side of the weld's heat-affected zone can crack. Four cracking sites should be considered for heat-affected zone cracking of the welds in the core shroud. In this case, the crack originates on the fuel side of the core shroud at the edge of the weld crown, the inside ring cracking site. Again, cracks are most likely to follow the angled weld profile. Weld crowns are variable, and some are not ground flush, which can limit the inspection surface.

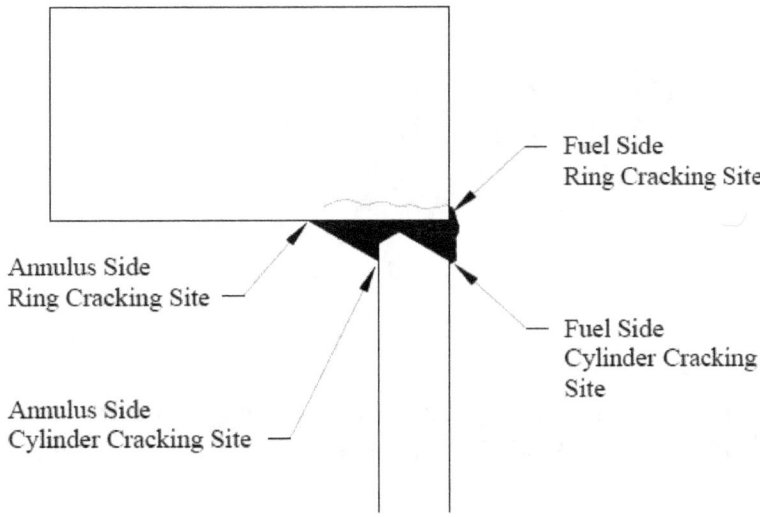

Figure 5-4 Four Crack Initiation Sites

5.2 Reflector Types

The mockup was designed to provide reflectors representative of field conditions and to meet the intent of the requirements of ASME Code Section XI, Appendix VIII performance demonstration test. At least 10 cracks are required over a range of depths with various conditions such as weld crown or limited scanning. Grading units are established with a minimum of 75 mm of weld, and blank grading units are provided as required. The types of flaws are to be representative of field flaws such as SCC.

Stress corrosion cracks are very difficult to grow to simulate the NDE responses of field-grown SCC. As a result, other crack types are used to mimic SCC for the purpose of evaluating NDE reliability. Two types of flaws were selected for the mockup—thermal fatigue cracks (TFCs) and weld solidification cracks (WSCs). These two crack types have very different ultrasonic response properties. The TFCs produce strong reflections similar to SCC that has become transgranular or fully open. The WSCs are reported to be more like tight SCC (Watson and Edwards 1996). The WSCs are made by filling an area that has been notched to the flaw dimensions (in depth and length) with a contaminated gas tungsten arc weld. The WSC forms in the welded notch.

Rough crack faces can be expected in the case of SCC. Figure 5-5 shows that thermal fatigue crack implants can provide such roughness. The shape and size of the crack can be measured before and after the implant in the mockup assemblies.

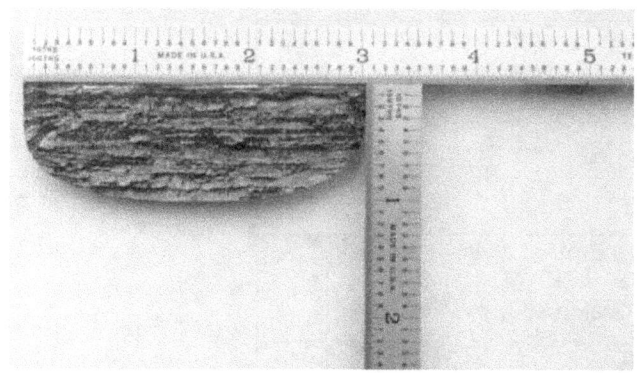

Figure 5-5   Example of Thermal Fatigue Crack Implant (photo courtesy of FlawTech, Inc., Concord, North Carolina; reprinted with permission)

Artificial TFCs are employed in welded stainless steel assemblies for tests of austenitic piping as specified in ASME Code Section XI, Appendix VIII. Figure 5-6 provides the notation and design of TFC implants. The mockup used this type of crack (and weld solidification cracks) to represent SCC in the core shroud. The top and bottom illustrations in Figure 5-6 are typical of piping in which the crack is formed after the weld preparation surfaces are machined onto the two pieces of stainless steel to be joined by welding but before welding starts. A tension bar is welded to one of the base metal pieces, and the TFC is initiated using a process of heating and cooling the desired crack location while the tension bar mechanically cycles the material. The crack forms, and the tension bar separates from the parent material with a small coupon of base metal attached. This coupon is removed from the tension bar and seal-welded back onto the base metal in preparation for welding. The welding is done, joining the two base metal pieces with the tight thermal fatigue crack created in base metal to create a representative condition for near-side inspection access. The idea is to create a crack that is oriented along the heat-affected zone of the weld and provide a sound path to the flaw face that is entirely within base metal.

The second illustration from the top in Figure 5-6 shows the work needed to create a TFC that is designed for inspection through the weld, far-side access. Here the weld is completed first, and then the crack is created in the heat-affected zone of the weld. To do this, an excavation that faces the weld is formed, and the tension bar is attached by welding. The thermal and mechanical cycling separates the tension bar from the test specimen with a piece of the test specimen attached to the tension bar as in the first case. Then the coupon is removed from the tension bar and seal-welded back into the test piece. The final step is to fill the excavation with weld metal to create the far-side inspection condition. This approach creates a crack that follows the orientation of the heat-affected zone and provides a sound path through the austenitic weld to the flaw face.

The third illustration in Figure 5-6 is a test of the acoustic properties of the implant process. The crack is formed with a tension bar, but the coupon is not replaced and seal-welded to form a crack. The area where the crack should be is repaired by welding. This approach creates a site for testing for detectable conditions associated with the coupon implant process.

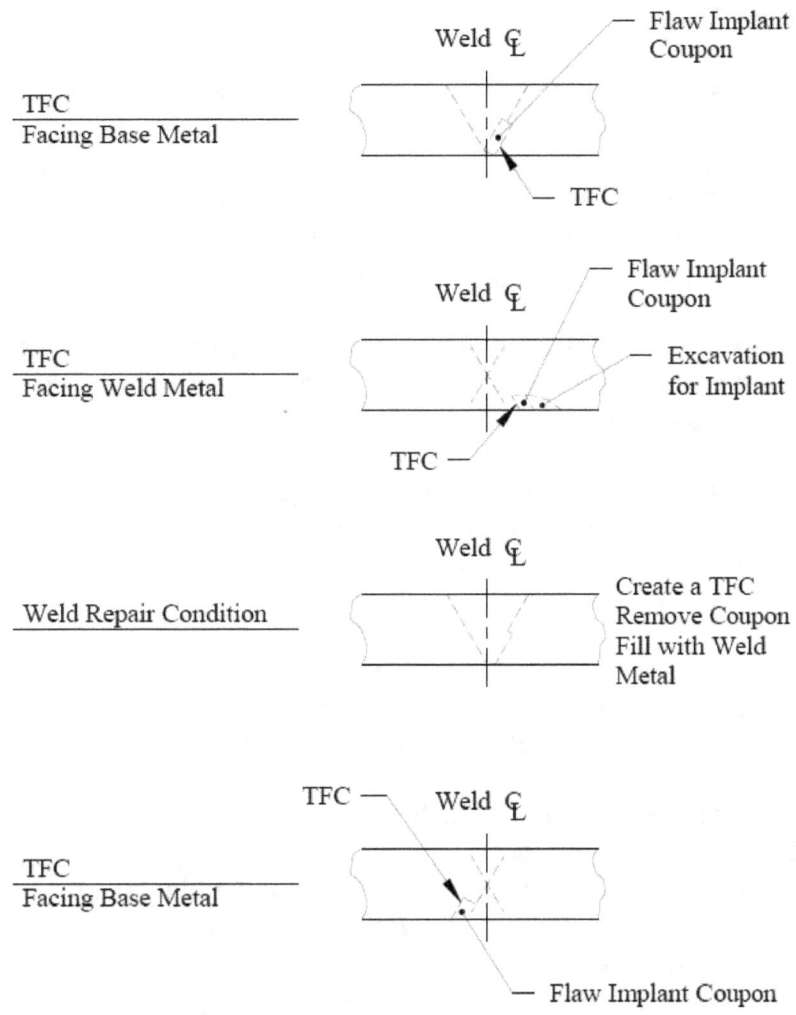

Figure 5-6 Thermal Fatigue Crack Notation

Machined reflectors in the PNNL mockup provided reference reflectors for calibration of the UT modalities that are used in the inspection of the core shroud. A number of notches were included for use in the laboratory phase of the study. These machined reflectors were used to establish the upper bound on the UT response and to quantify response dependence on reflector properties such as size and tilt. Notches were used in an evaluation of material properties such as attenuation and scattering. A series of vertical notches, from 5% to 75% through-wall, and notches at positive and negative angles relative to vertical at 30% through-wall, were built for the parametric study.

The mockup was designed to permit evaluation of UT reliability for inspection through base metal and weld metal. All cracks and notches were installed in the heat-affected zone of the weld. Because the TFCs are implanted with weld metal on one side, the mockup contained separate flaws for assessing base metal and weld metal inspection effectiveness.

5.3 Acceptance Tests

It was important to measure some of the responses for ultrasonic modes being used by industry before completing the mockup design. The data from the mockup were sufficient to estimate flaw response as a function of flaw through-wall extent for most of the UT inspection modalities described in Section 3. The responses were estimated separately for corner-trap and tip-diffracted signals. The responses and their variances were estimated separately for all inspection modalities used to inspect through the weld.

As described in this section, there are a number of cases in which the most convenient inspection is through the weld metal. The reliability of these inspections is known to be in doubt and, for this reason, it was estimated separately.

Sometimes detection could be accomplished using diffracted signals. This detection method was selected when no corner-trap method was available. Diffracted signals are generally 20 dB lower in amplitude than corner-trap reflections, so the use of diffracted signals was evaluated separately in terms of their signal-to-noise ratio.

The grain structure of stainless steel limited the UT response, and higher frequencies were often strongly attenuated, especially in thicker material. Variability in grain structure, in different samples of type 304 stainless steel, had an impact on ultrasonic signal strengths.

Table 5-2 shows the specimens used to validate the mockup design, a necessary first step before building the complete mockup for use in a parametric study and a blind test estimating ISI performance. Two types of welding were used—shielded metal arc and submerged arc welding. Four weld profiles were used—single and double V, J, and K. Each of the first five plates shown in Table 5-2 was inspected and UT images created. A signal-to-noise level for each of the notches and TFCs was calculated from the UT images and formed the basis for evaluation of the four weld types (single V, double V, J, and K).

Table 5-2 First Five Plates Used in Design Tests for Mockup

Plate Name	Weld Type	Calibration	Cracks	Notches
CS-02-01	Shielded metal arc, Single V	3 side drilled holes	10% TFC	10% and 20%
CS-02-02	Submerged arc, Single V	3 side drilled holes	10% TFC	10% and 20%
CS-02-03	Shielded metal arc, Double V	3 side drilled holes	None	10% and 20%
CS-02-04	Shielded metal arc, J weld	3 side drilled holes	None	10% and 20%
CS-02-05	Shielded metal arc, K weld	3 side drilled holes	None	10% and 20%

A signal-to-noise ratio of 6 dB or greater was required for a flaw to be recorded. This is based on prior round robin test experience in which detection results were found to be unreliable when the signal-to-noise ratio was less than 6 dB. In the base metal inspection, the creeping wave mode detected all of the flaws with a signal-to-noise ratio of 16 dB or greater. Turning to the difficult inspection of weld metal, there was no detection of the 10% notch in either the double-V or the single-V weld. The 10% notch in the K weld, as viewed from the angled-side, also was undetected. All three of these welds produced a difficult-to-inspect condition.

A 40% through-wall weld repair condition was fabricated. First, a TFC that followed the 30° angle of the weld fusion line was created. This 6-mm thick, 40% through-wall coupon was removed and the double-V weld and repair made with the shielded metal arc welding process. This weld repair was imaged with the 2-MHz dual, 10×34 mm, 50-mm focus, 75° longitudinal probe, from the four possible access directions with results as follows.

For inspections from the scan surface on the same side of the assembly as the repair with weld metal access, the weld repair was not acoustically detected. For inspection from the repaired surface with base material access, the weld signals in the repair region were higher in amplitude than outside the weld repair area. The peak signals from the weld repair weld signal areas were −8.2 and −7.1 dB, respectively. The base metal peak weld signals were −9.5 dB. Thus we see a difference of up to 2.4 dB in the weld repair region. This difference is small enough that it would be called a no-detect condition. Images from the opposite surface inspection show no detection of the weld repair from either the base metal or weld metal access.

Figure 5-7 shows the responses from notches, side-drilled holes, and thermal fatigue cracks using 45° shear waves. Ultrasonic inspections of the submerged arc single-V welded plate were performed. This plate contains one 10% deep and one 20% deep notch, a 10% deep TFC, and three side-drilled holes at one-quarter, one-half, and three-quarters of the wall thickness depth. Data in Figure 5-7 show that the base metal inspection is straightforward. The strength of this mode is that a strong corner-trap signal for flaw detection and a weaker tip signal for flaw depth sizing are found in base metal inspection. The confusion from multiple wave modalities is eliminated, as the compression (longitudinal) wave is not present. This wave, however, did not penetrate the weld metal very well.

The responses in Figure 5-7 are given relative to the noise level. The root mean square noise level is shown in decibels as a function of time from the transducer excitation. Shoe noise is the largest contributor to noise level until approximately 30 microseconds has passed. Part noise, from the stainless steel, is shown at about −77 dB. An elevated noise level at the back surface of the component is shown.

Figure 5-8 provides the responses using 45° shear waves for base metal and weld metal access. Base metal results (hollow symbols) show that all defects were detected. This includes the notch tip signals that are 25 dB below the corner-trap signals (the TFC data are not plotted but fall approximately 5 dB lower than notch data). The weld metal inspection detected only the one-quarter wall thickness deep side-drilled hole and both notch corner trap signals. These signals are significantly lower than their base metal counterparts (−15 dB and −35 dB, respectively).

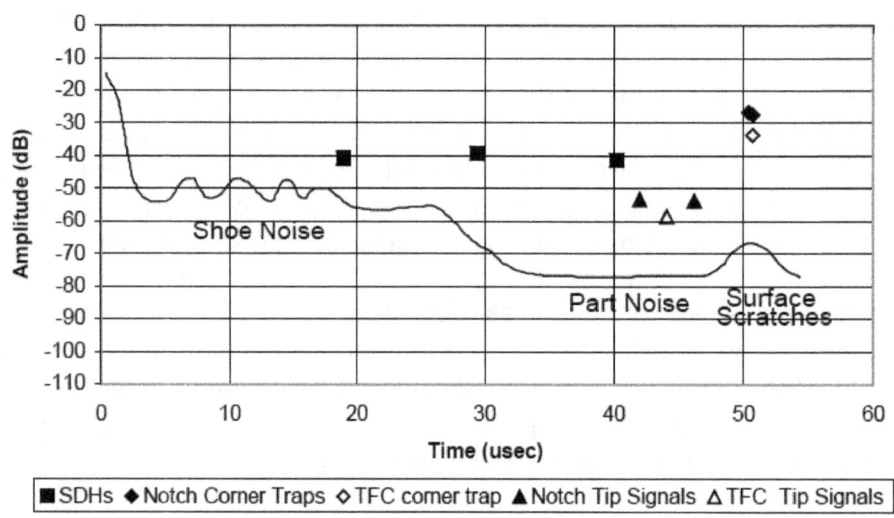

Figure 5-7 **Near Side Access (Through Base Metal) Responses from Notches, Side-Drilled Holes, and Thermal Fatigue Cracks Using 45° Shear Waves in a Submerged Arc Single-V Welded Assembly.**

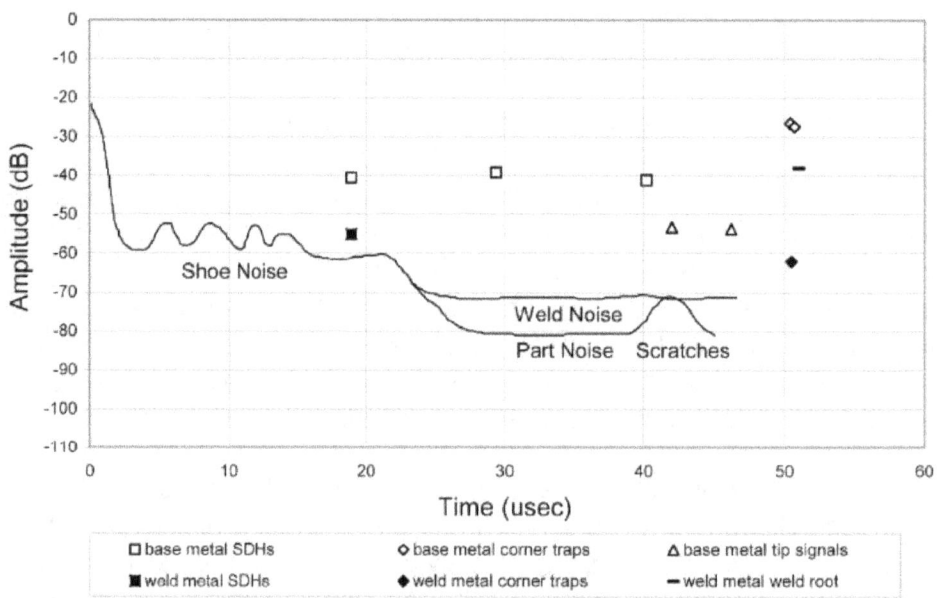

Figure 5-8 **Responses Using 45° Shear for Base Metal and Weld Metal Access in a Submerged Arc Single-V Welded Assembly**

Data from the 60° longitudinal, 2-MHz, 75-mm focus probe showed that the base metal inspection found the three side-drilled holes (SDHs), the flaw corner-trap, and tip signals. From the weld metal inspection only the one-quarter wall thickness and the three-quarters wall thickness SDH signals were found. Results are shown in Figure 5-9.

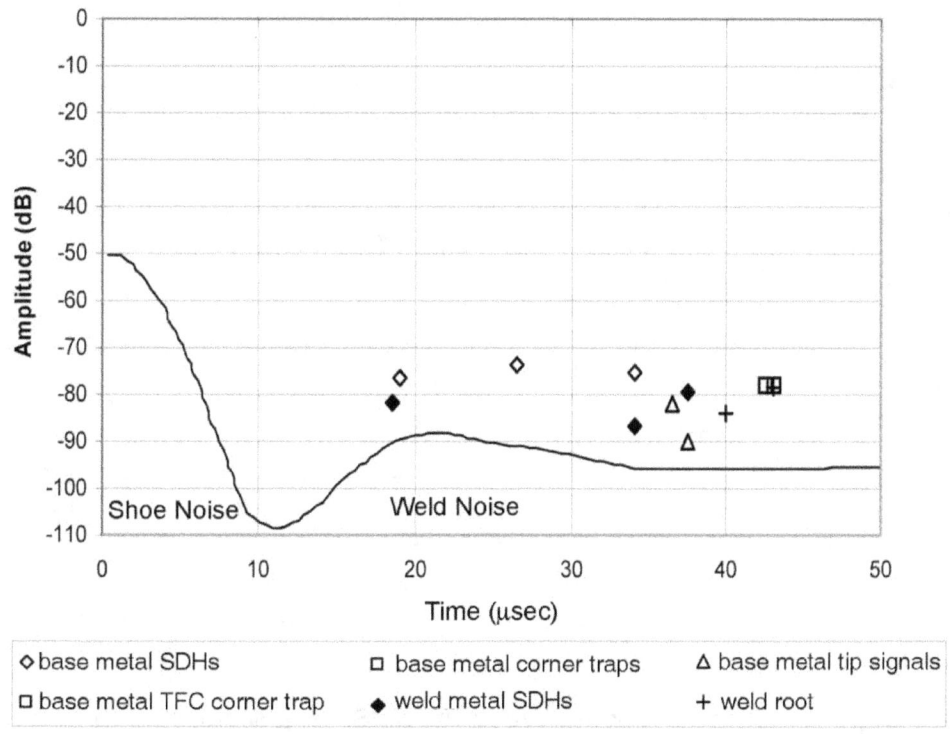

Figure 5-9   Responses Using 60° Longitudinal for Base Metal and Weld Metal Access in a Submerged Arc Single-V Welded Assembly

Of the probes evaluated in this laboratory study, the probe having the best through-weld inspection effectiveness was the creeping wave probe. At the design angle of 75°, a longitudinal wave is produced along with a 33° shear wave, and a creeping wave (with longitudinal velocity) that travels just below the surface is also generated. Four modes of inspection are possible with this probe as was shown in Figure 3-5. The first mode is from the near-surface creeping wave that will detect very deep cracks or crack openings (base of cracks) that are scan-surface-connected. The second mode is the far-surface creeping wave generated by transverse (head) waves leaking from the near-surface creeping waves, which convert to a creeping wave at the far surface. The third mode is from the 33° shear wave mode converting at the far surface to a longitudinal wave then reflecting off a crack face and returning to the transducer as a longitudinal wave (noted as the SH-L-L wave). The fourth mode is the 75° longitudinal wave mode.

The results from the SH-L-L mode are shown in Figure 5-10. Both base metal and weld metal access conditions show detection of the two notches and one TFC. The response levels differ in that the base metal 20% notch is approximately 16 dB higher than the 10% TFC or 10% notch. In contrast, the weld metal shows only a 5 dB to 9 dB difference with the 10% TFC actually appearing stronger than the 10% notch. However, in both weld metal and base metal inspections, the 20% notch response is larger than the 10% flaw responses, as expected.

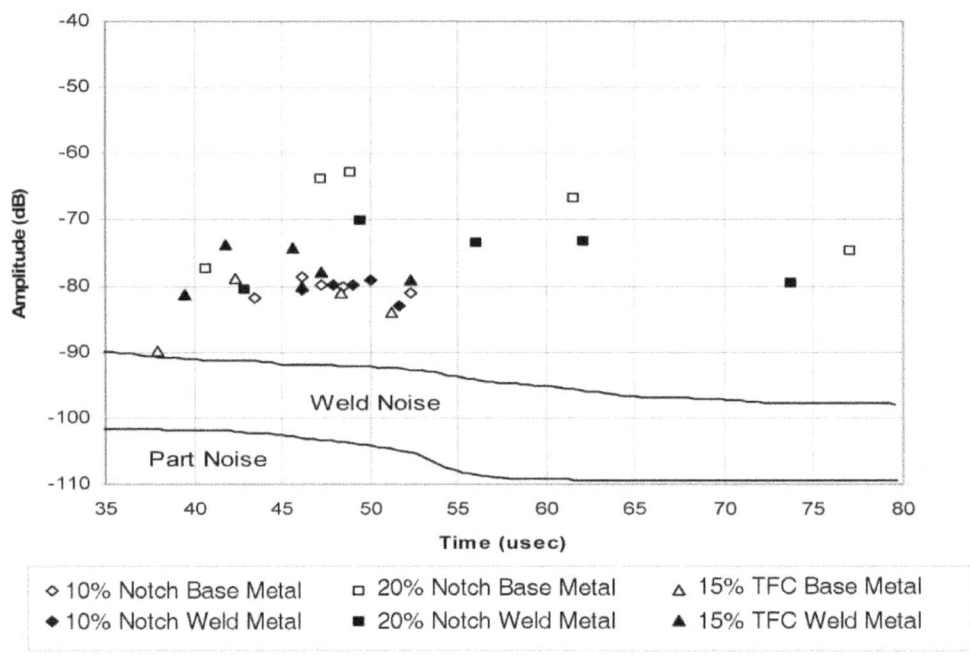

Figure 5-10   Response Using Creeping Wave Probe SH-L-L Mode in a Submerged Arc Single-V Welded Assembly

Figure 5-11 shows the responses for the 75° compression (longitudinal) mode of a creeping wave probe. From the manual results (confirmed also by automated data), the 10% TFC from the weld access side is the strongest signal. It is 5 dB stronger than the base metal access condition and occurs earlier in time. The 20% notch produces the same response from both base metal and weld metal access conditions, but the weld metal response is delayed in time due to velocity differences in the weld material or beam steering. Similar results are found for the 10% notch.

Table 5-3 gives the test results used for guiding the mockup design. Responses are measured from corner traps, side-drilled holes, and crack tips. Four ultrasonic modes are used—45° shear, 60° longitudinal, creeping wave (SH-L-L mode), and 75° longitudinal. Two access conditions were tested—cracks on the near side of the weld and cracks on the far side of the weld. Table 5-4 lists the far-side access responses from a 20% notch for 75° longitudinal probe for the various weld profiles. The double-V weld profile had the lowest response of all the weld profiles for far-side access using high-angle longitudinal waves. From this data, the double-V weld was selected as the difficult configuration for the mockup. Two K-weld TFCs are also in the mockup and will allow a comparison of TFCs in the K and double-V welds.

In summary, the inspection through base metal is straightforward. Flaw tip signals were found with the 45° shear, 45° longitudinal, 60° longitudinal, and 75° longitudinal probes, thus providing sizing information. Signal-to-noise levels of approximately 8 decibel or better were achieved. Inspections through the weld metal, however, proved to be more difficult. Detection of the shallow flaws is accomplished most reliably with the 75° longitudinal probe. All three reflectors (10% and 20% notches, and 15% TFC) were detected with each of the three probe modes— longitudinal wave, shear-L-L wave, and ID creeping wave.

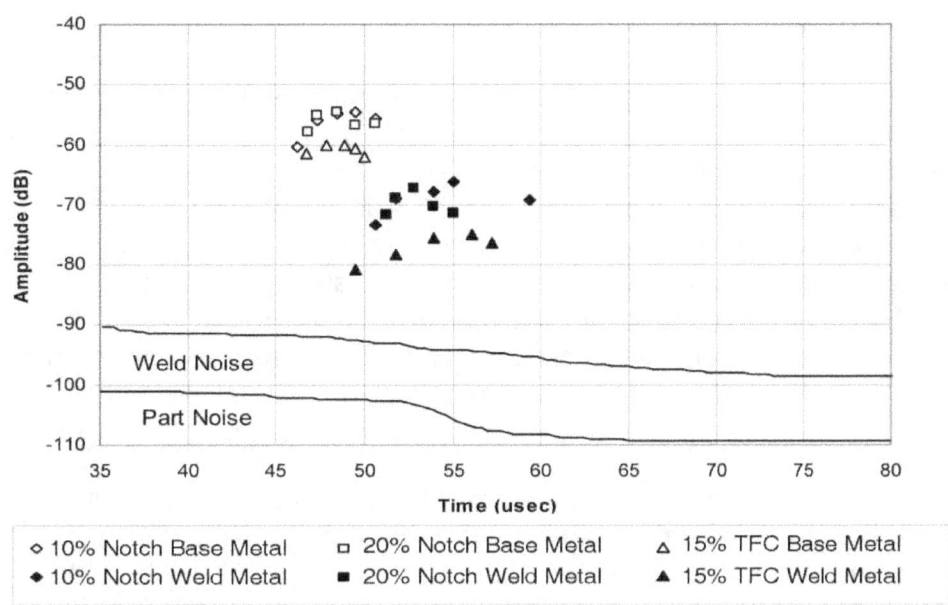

Figure 5-11 Responses Using Creeping Wave Probe 75° Longitudinal Mode in a Submerged Arc Single-V Welded Assembly

Table 5-3 Test Results Used for Mockup Design – Submerged Arc Single-V Welded Assembly

Ultrasonic Mode	Corner Trap	Side Drilled Holes	Tip Signal
45° shear, near-side	−26 dB	−40 dB	−53 dB
45° shear, far-side	−62 dB	−55 dB	No detect
60° longitudinal, near-side	−78 dB	−75 dB	−82 dB
60° longitudinal, far-side	No detect	−82 dB	No detect
Creeping wave, SH-L-L, near-side	−62 dB		
Creeping wave, SH-L-L, far-side	−70 dB		
75° longitudinal, near-side	−55 dB		
75° longitudinal, far-side	−68 dB		

Table 5-4 Far-Side Access Responses from a 20% Notch for 75° Compression (Longitudinal) Waves for Various Weld Profiles

Weld Profile	75° L Response, Far-Side Access
Single V, SMAW	−70 dB
Single V, SAW	−68 dB
Double V	−76 dB
J	−75 dB
K	−72 dB

5.4 Completed Mockup

The mockup consists of small welded plates designed to represent conditions important to ultrasonic inspection of the core shroud and the shroud support assembly. These small welded plates represent the vertical and horizontal shroud weld cracking sites. Notches are included in the mockup for an evaluation of the effects of through-wall extent and tilt. Thermal fatigue cracks and weld solidification cracks have been placed in the heat-affected zones of the weld in sufficient number so that the mockup can be used to simulate a blind test that meets the intent of ASME Code Section XI, Appendix VIII. The mockup can be divided into "grading units" with twice the amount of blank material as cracked material. The assemblies permit inspection from both sides (fuel and annulus sides) and weld crowns include both "as-welded" and ground-flush conditions.

The mockup contains both 304 stainless steel and Alloy 600 (Inconel). The Alloy 600 material costs, machining costs, and welding costs are two or three times higher than for type 304 stainless steel, so the mockup uses relatively little Alloy 600. The Alloy 600 portion of the mockup is intended for comparing and contrasting the ultrasonic signal qualities with more extensive measurements in 304 stainless steel. The mockup consists of 40 welded assemblies of 304 stainless steel and 10 welded assemblies of Alloy 600.

The mockup contains representative weld profiles because they are likely to affect the UT response. The weld profile changes the grain alignment of the weld metal because the adjacent base metal serves as the principal heat sink. The amount of heat generated in the welding process is greater for submerged metal arc welding than for manual welding. The heat deposition rate and the size of the weld bead affect the attenuation and scattering properties of the weld metal.

The mockup permits evaluation of UT reliability for inspection through base metal and weld metal. Thermal fatigue cracks are installed in the heat-affected zone of the weld. Because the thermal fatigue cracks are implanted with weld metal on one side, the recommended mockup contains separate flaws for base metal and weld metal inspection.

The mockup was designed to permit inspection on both sides of the assemblies. In an inspection of the core shroud, cracks can be connected to the scanning surface or connected to surface opposite to the scanning surface. The mockup also permits simulation of inspection from the inside and outside of the core shroud.

Machined reflectors are provided for calibration of the UT modalities used in core shroud inspections. A number of notches are included for use in the laboratory phase of this study. These machined reflectors will be used to establish the upper bound on the UT response and to quantify the response's dependence on reflector properties such as size and tilt. The notches will be used also in an evaluation of material properties such as attenuation and scattering. Table 5-5 shows the notch depths and angles included in the mockup.

Table 5-5 Notch Depths and Angles Selected for Mockup

Depth (%)	Angle (degrees)									
	−30	−25	−20	−15	−10	−5	0	5	15	25
5							x			
10							x			
20							x			
30	x	x	x	x	x	x	x	x	x	x
40							x			
75							x			

The mockup was designed within the spirit of the ASME Code Section XI, Appendix VIII. At least ten cracks are provided over a range of depths with various conditions such as weld crown or limited scanning. At least four of the cracks are 10% to 30% through-wall, at least four of the cracks are 31% to 60% through-wall, and at least two of the cracks are 61% to 100% through-wall. The cracks are randomly located in grading units. Grading units are described in Section XI, Appendix XIII of the ASME Code.

Grading units were established with a minimum of 75 mm of weld, and blank grading units were provided as required. The grading units were all the same size with one or two grading units located on each side of the weld for every assembly. Grading units were located randomly along the weld, and cracks were located within the grading units according to the ASME Code. Enough flaw-free material was provided so that 20 blank grading units were contained in the mockup.

The mockup consisted of both stainless steel plates and Alloy 600 plates, 32 welded assemblies of 304 stainless steel and 5 welded assemblies of Alloy 600. A complete list of mockup assemblies appears in Table 5-6.

Table 5-6   Mockup Assemblies. Size of the cracks is withheld to permit mockup re-use in future blind tests.

Material, Thickness	Weld	Surface	Reflectors
304 SS, 50 mm	SMAW, single V	Flush	TFC, 10 and 20% notches, 3 SDHs
304 SS, 50 mm	SAW, single V	Flush	TFC, 10 and 20% notches, 3 SDHs
304 SS, 50 mm	SMAW, double V	Flush	10 and 20% notches, 3 SDHs
304 SS, 50 mm	SMAW, J	Flush	10 and 20% notches, 3 SDHs
304 SS, 50 mm	SMAW, K	Flush	10 and 20% notches, 3 SDHs on flat side, 10 and 20% notches on angled side
304 SS, 50 mm	SMAW, double V	Flush	Base metal facing TFC, weld repair, weld metal facing TFC
304 SS, 50 mm	SMAW, double V	Flush	2 TFCs
304 SS, 50 mm	SMAW, double V	Flush	2 TFCs
304 SS, 50 mm	SMAW, double V	Flush	Blank
304 SS, 50 mm	SMAW, double V	Flush	Blank
304 SS, 50 mm	SMAW, double V	Flush	Blank
304 SS, 50 mm	SMAW, double V	Flush	Blank
304 SS, 50 mm	SMAW, double V	Flush	Blank
304 SS, 50 mm	SMAW, double V	As welded	Blank
304 SS, 50 mm	SMAW, double V	Flush	Blank weld, 10% end milled notch
304 SS, 50 mm	SMAW, double V	Flush	30% and 40% notches
304 SS, 50 mm	SMAW, double V	Flush	5% and 75% notches
304 SS, 50 mm	SMAW, double V	As welded	30% notch
304 SS, 50 mm	SMAW, double V	Flush	30% notches at −5, −10, 15°
304 SS, 50 mm	SMAW, double V	Flush	30% notches at −20, −25, −30°
304 SS, 50 mm	SMAW, double V	Flush	30% notches at 5, 15 25°
304 SS, 50 mm	SMAW, double V	Flush	TFC
304 SS, 50 mm	SMAW, double V	Flush	2 TFCs
304 SS, 50 mm	SMAW, double V	As welded	TFC
304 SS, 50 mm	SMAW, double V	As welded	TFC
Alloy 600, 75 mm	SMAW, double V	Flush	TFC
Alloy 600, 75 mm	SMAW, double V	Flush	TFC
Alloy 600, 75 mm	SMAW, double V	Flush	TFC
Alloy 600, 75 mm	SMAW, double V	Flush	Blank weld, 10% end milled notch
304 SS, 50 mm	SMAW, double V	Flush	50% notch
304 SS, 50 mm	SMAW, double V	Flush	TFC
304 SS, 50 mm	SMAW, double V	As welded	TFC
304 SS, 50 mm	SMAW, K	Flush	2 TFCs
Alloy 600, 75 mm	SMAW, double V	Flush	TFC
304 SS, 50 mm	SMAW, double V	Flush	WSC
304 SS, 50 mm	SMAW, double V	Flush	WSC
304 SS, 50 mm	SMAW, double V	Flush	WSC

SS: stainless steel; SMAW: shielded metal arc weld; SAW: submerged arc weld; TFC: thermal fatigue crack; WSC: weld solidification crack; SDH: side-drilled hole

6 PARAMETRIC STUDY

An analysis of signal-to-noise ratio is important for all physical measurements, especially where performance is relatively low. The effects of surface finish, crack-to-crack variability, and other factors that cause the responses to vary should be qualified to predict reliability and an analysis of variance is one method to achieve this.

Figure 6-1 provides an example of inspection noise sources with variance. Three noise sources are shown—base metal part noise, shoe noise, and electronic noise. The responses are shown using a logarithmic scale and are calibrated using a standard reflector—a 10% end milled notch in a 50-mm thick stainless steel specimen. Time in the part is shown in microseconds on the abscissa, and this scale can be converted to part path by multiplying by the speed of sound in the material and dividing by two.

Electronic noise is shown at a constant level of approximately −100 dB, five orders of magnitude below the response from the reference reflector. The vertical bars on the data points show one standard deviation, approximately ±2 dB, with the responses from electronic noise. Electronic noise can be reduced by improving the electronics or by temporal averaging.

Shoe noise is caused by acoustic echoes within the standoff material, which for contact probes is typically a plastic wedge. Shoe noise, the principal noise source early in time, decays to the level of the electronic noise late in time. Shoe noise is reduced by separating the receiver from the transmitter in the standoff material and in optimizing the plastic wedge design to reduce acoustic reflections in the wedge.

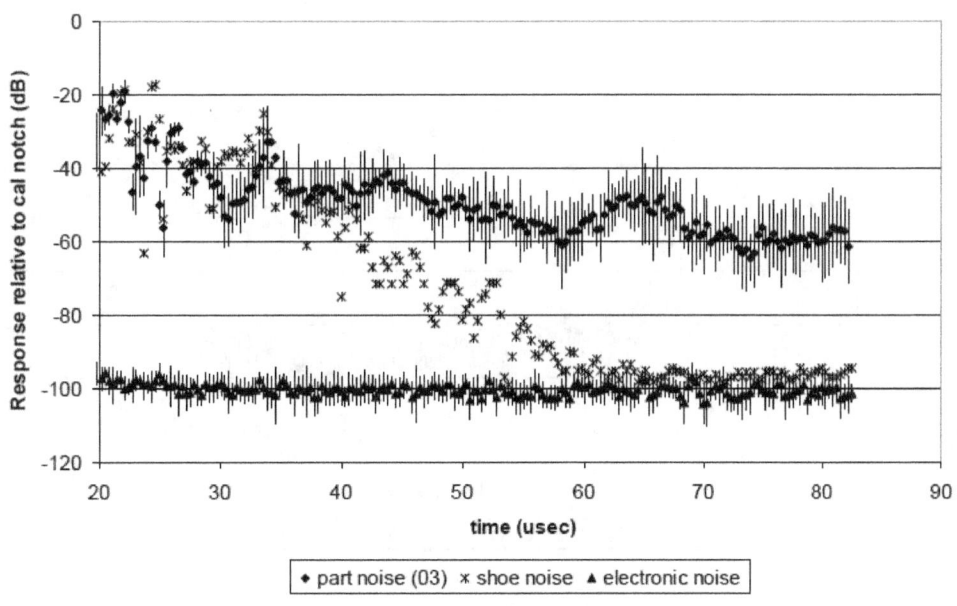

Figure 6-1 Noise Sources Using 45° Shear Waves

Part noise is caused by the microstructure of the base metal, type 304 stainless steel. This is the principal noise source for inspections of stainless steel components. Part noise levels vary but in the example drop by −40 dB after 70 microseconds, three orders of magnitude below the calibration reflector. Part noise can be reduced by using spatial averaging techniques.

6.1   Near-Side Access

Measurement of detection performance in the parametric study is useful for predicting performance of an inspection in the field and predicting the results of a screening test. Performance is measured by the ability to correctly distinguish cracked from blank material. The calibration reference reflector that was used for these tests was a 10% through-wall notch.

6.1.1   45° Shear Waves at 2.25 MHz, Base Metal Access

Figure 6-2 shows the response-time chart of a highly effective detection test, the response of notches insonified through base metal at 2.25-MHz, 45° shear waves. The high performance is evidenced by the separation of the corner-trap and tip signal from the non-flaw interference and noise sources. These data were obtained with the notches on the surface opposite to the inspection surface. Figure 6-2 shows the part noise with standard deviation, interfering signals from surface scratches, interfering signal from weld noise, corner trap signals, and tip signals. The nomenclature used in this figure is that a notch that has a through-wall depth of 75% will be identified as N75.

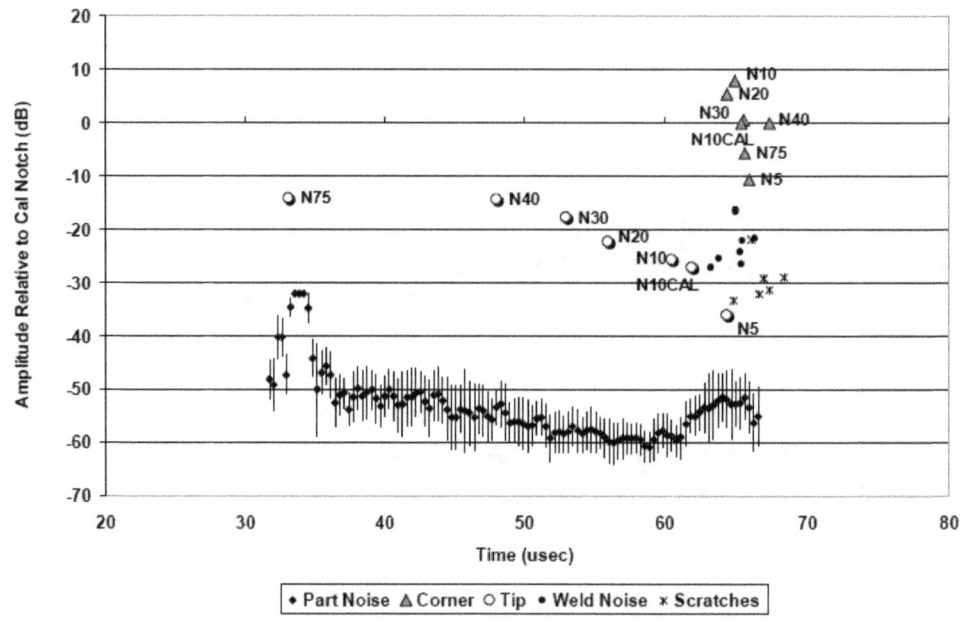

Figure 6-2   Responses of Notches to 45° Shear Wave Insonified Through Stainless Steel Base Metal

6-2

The part noise is shown as a time series of mean and standard deviation of 16 A-scans taken on stainless steel mockup plate No. 03. This noise level is present in the base material and is produced by backscattered ultrasonic energy from the base metal microstructure.

Data for surface scratches are also shown in Figure 6-2, and they produce a response of about 20 dB higher than the minimal part noise. Time of flight to the scratches is not different from the corner trap signals from the notches. Signals from weld microstructure (weld noise) produce response and time behavior similar to scratches.

Corner-trap responses from the notches are +40 dB to +60 dB above the noise floor and +6 dB to +26 dB above the scratches and weld noise. The amplitude variability of 20 dB in the notches cannot be explained by notch size, as noted in Section 6.2. Tip signals are also shown in Figure 6-2. Good amplitude-time separation from the other signals is shown except for the tip signal from the 5% notch.

6.1.2   75° Longitudinal Waves at 2.0 MHz, Base Metal Access

Figure 6-3 shows the results of the 75° longitudinal wave at 2.0 MHz on the notches in the mockup specimens. These data were obtained with the notches on the surface opposite to the inspection surface. Figure 6-3 shows the part noise with standard deviation, corner trap signals, and tip signals.

The part noise is shown as a time series of mean and standard deviation of 16 A-scans taken on stainless steel mockup plate #23. This noise level is present in the base material and is produced by backscattered ultrasonic energy from the base metal microstructure. The noise level is elevated with respect to the calibration notch (−25 dB) compared to the 45° shear-wave (−55 dB). Note: the scattering has not increased relative to the shear-waves; instead, the calibration-notch response is diminished by mode conversion losses.

Corner trap responses from the notches are −6 dB to +10 dB relative to the calibration notch. The amplitude variability of 16 dB in the response from the notches does not follow the same pattern as the 45° shear-wave data. Considerable time variability is also evident.

Tip signals are also shown in Figure 6-3. Good amplitude-time dependence is shown except for the tip signal from the 10% notch. Most of the tip signals are at 25 dB to 35 dB of the noise level because they are not diminished by mode conversion losses.

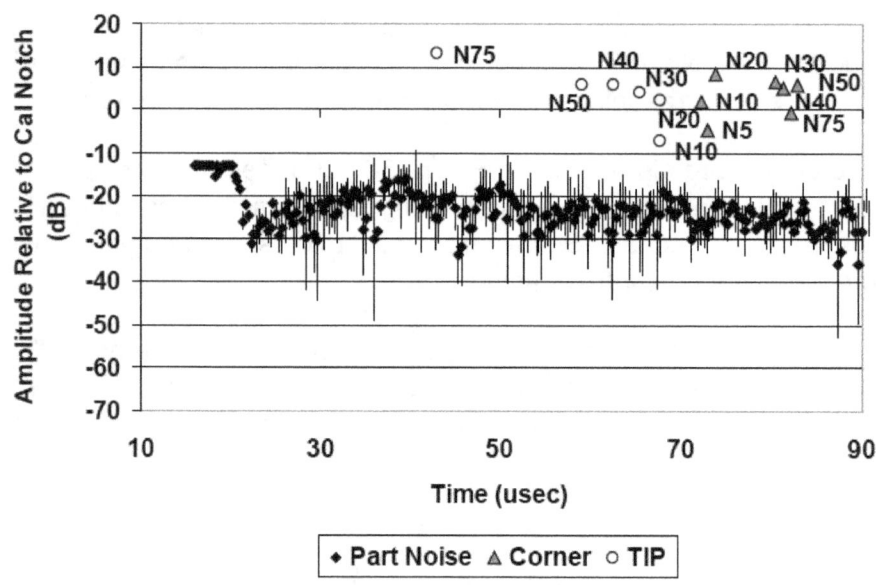

Figure 6-3   Response of Notches to 75° Longitudinal Waves at 2.0 MHz Insonified Through Stainless Steel Base Metal

6.1.3   Crack Responses

The responses from the TFCs and WSCs were measured when insonified through base-metal using 45° shear waves. Figure 6-4 shows the TFC responses are −6 dB of the notches (see Figure 6-2). The signals from the WSCs could be found with careful measurement and a-priori knowledge of crack location. But the signals from the WSCs could not be distinguished from all of the interference signals. Figure 6-5 shows the WSC responses for the 30% through-wall crack are −20 dB of the response from a 30% through-wall TFC when using 45° shear waves.

6.1.4   Thermal Fatigue Cracks in Alloy 600

Figure 6-6 shows the response-time chart for thermal fatigue cracks in Alloy 600 using 45°, 2.25-MHz SV-waves for base metal access. The response from the cracks varied from −8 dB to −22 dB of the calibration notch. Some interference from the weld was present at −21 dB. Surface scratches produced interference at a relatively low level of −30 dB. Part noise is shown with standard deviation. The responses from the tip signals of the thermal fatigue cracks are shown to be clearly separate from the part noise.

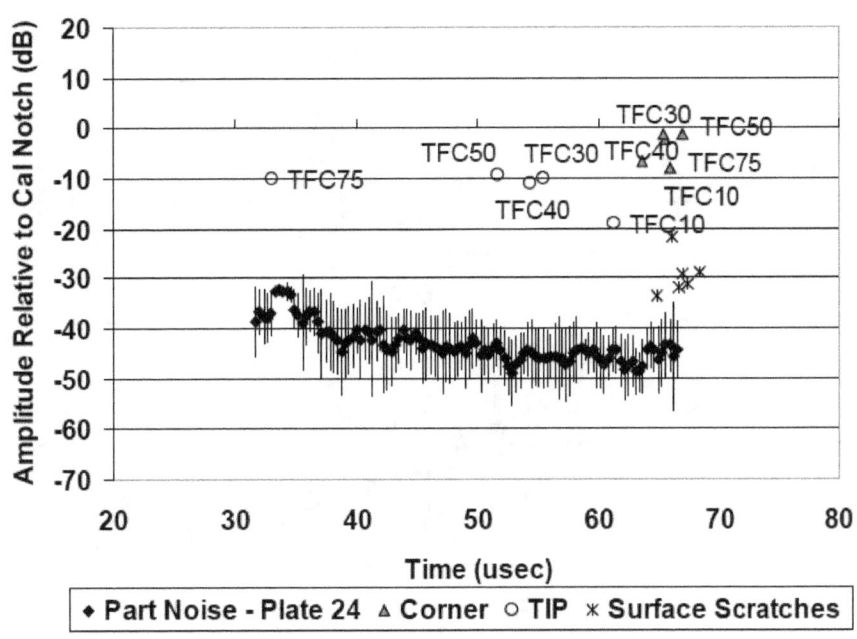

Figure 6-4   Thermal Fatigue Crack Responses to 45° Shear Waves at 2.25 MHz Insonified Through Stainless Steel Base Metal

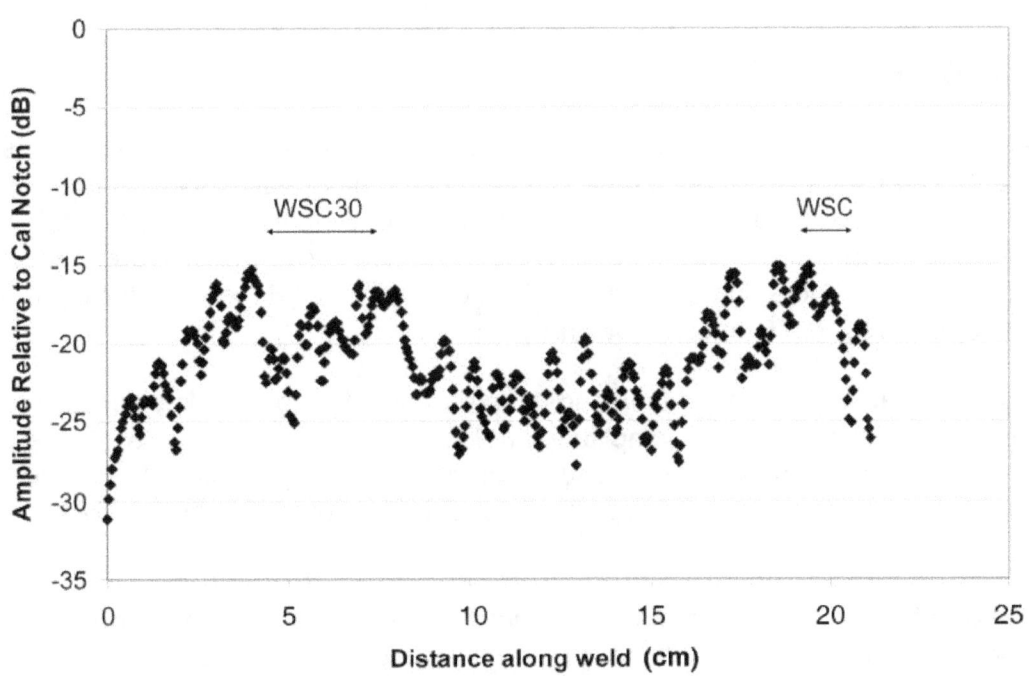

Figure 6-5   Weld Solidification Cracks on the Near Side Using 45° Shear Waves in Stainless Steel

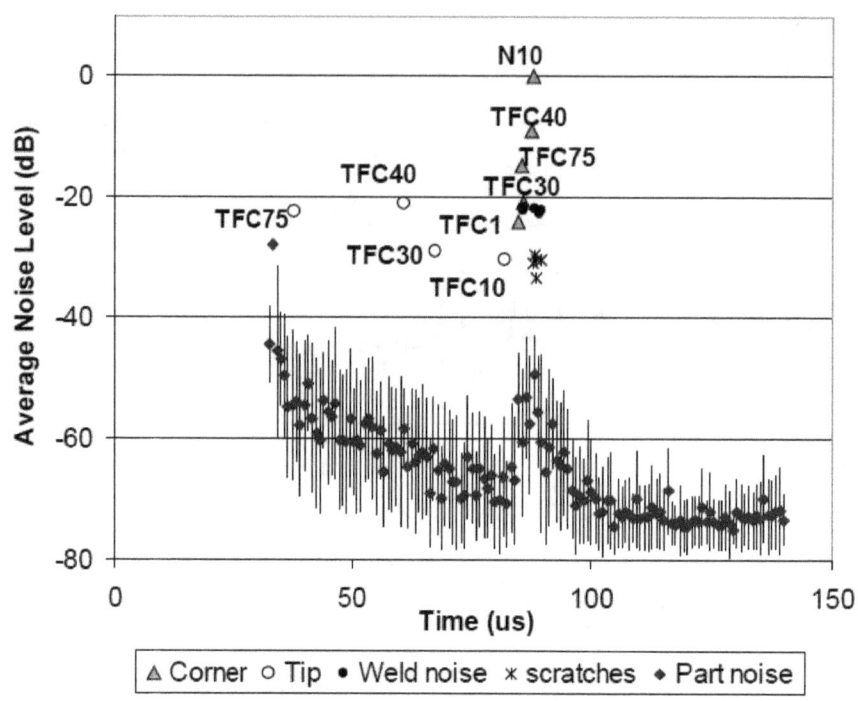

Figure 6-6 Response of Thermal Fatigue Cracks for Near-Side Access in Alloy 600

6.2 Echo Dynamics

The echo-dynamic curves were recorded for the response of notches to 45° shear waves in stainless steel. The analysis described here is applicable to other cases, but the specific results will vary with part thickness, notch size, and transducer characteristics. Figure 6-7 shows the echo-dynamic curves for three of the notches in the mockup. These curves were measured to the surrounding noise level. Note that the shapes of the curves do not depend on through-wall extent from a 5% (N5) to a 75% (N75) through-wall notch.

Figure 6-8 shows that the duration of the echo-dynamic curve for notches in stainless steel has an approximately linear relationship with amplitude (in decibels). No relationship exists between notch size and the echo-dynamic curve because, for the transducer selected, the amplitude of the response determines all of the properties of the curves, and amplitude does not depend on notch size.

Figure 6-9 shows the amplitude response of the notches as a function of their through-wall extent. The data shown in Figure 6-9 were taken with 45° shear, 9.5-mm-diameter transducers. The notches were on the opposite surface as in Figure 6-2. These data were taken to assist in explaining the unexpected variability in response from the notches. The notches were made part of the mockup design because they should produce simple, easily explained responses. In Figure 6-9, the 1.5-MHz transducer produced the expected response. The 10% notch filled the transducer beam, and the larger notches produced a response equivalent to the 10% notch. The 5% notch did not fill the transducer beam and consequently gave a reduced response.

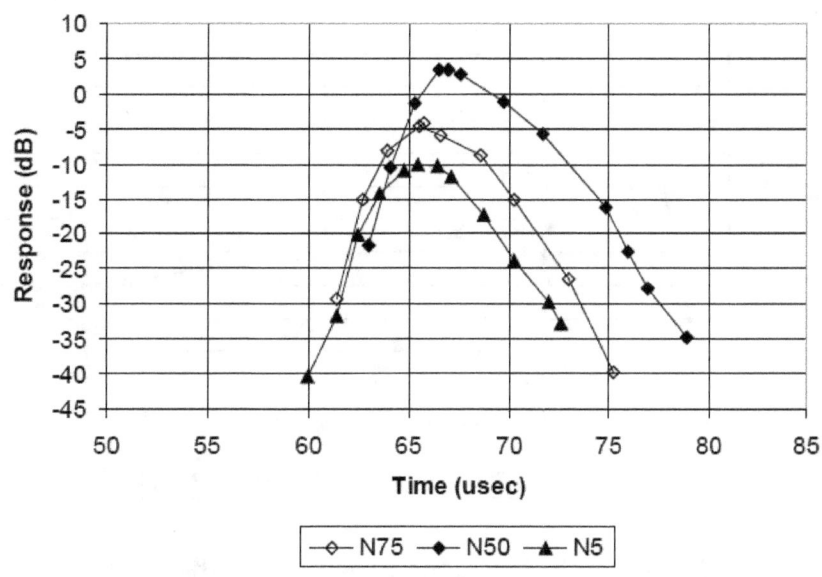

Figure 6-7 Echo-Dynamic Curves for Three Notches in Stainless Steel

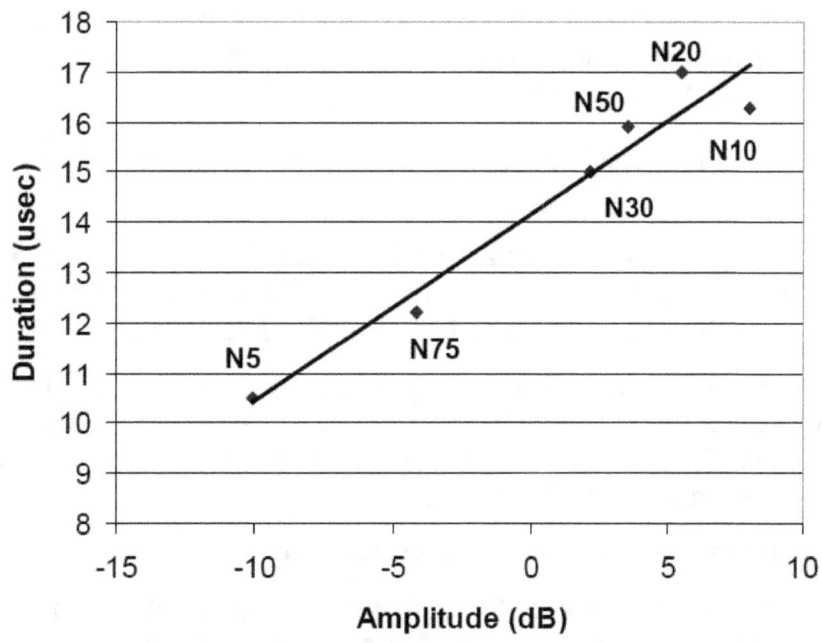

Figure 6-8   Amplitude versus Duration of the Echo-Dynamic Curve for Notches in Stainless Steel

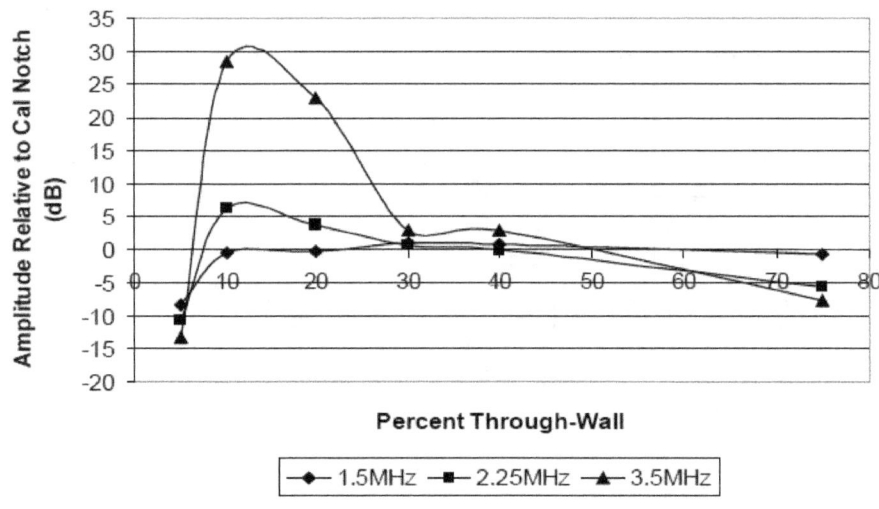

Figure 6-9   Response of Notches in Stainless Steel to 45° Shear Waves at Three Frequencies

The 2.25-MHz data are taken from Figure 6-2. The 10% and 20% notches are 4 dB to 6 dB above the response from the calibration notch (10%) when they should be 0 dB. The 75% notch is −5 dB of the calibration notch. The 3.5-MHz data show that the response variations are systematically amplified at the increased frequency.

A combination of effects is probably producing the variations. The principal effects that were considered are

- *surface finish* – The specimen that contains the 10% and 20% notches has a machined-surface finish. The other specimens have an "as-rolled" finish. This is considered the principal cause of the increased response in the 10% and 20% notches.

- *base metal variability* – The part noise in the specimens varies by 6 dB with the specimen that contains the 10% and 20% notches showing the least scattering and the specimen that contains the 5% and 75% notches showing the most. This is considered the principal cause of the decreased response in the 75% notch.

- *notch tilt and roughness error* – This is not considered to be the principal cause.

- *notch length* – The notches were specified with a 2-to-1 aspect ratio (length-to-depth). It is possible, but unlikely, that the notch length affects the coherency of the energy that is returned to the transducer.

The echo-dynamic curves were recorded for the response of thermal fatigue cracks to 45° shear waves. Figure 6-10 shows the echo-dynamic curves for two thermal fatigue cracks and the calibration notch in the Alloy 600 specimens. These curves were measured to the surrounding noise level. The TFCs are listed in order of increasing through-wall extent, with TFC 3 the smallest and TFC 4 the largest. The duration of these echo-dynamic curves showed some dependence on crack size. The difference in smoothness between notches and cracks accounts for this difference in dependence of the duration of the echo-dynamic curve on reflector size.

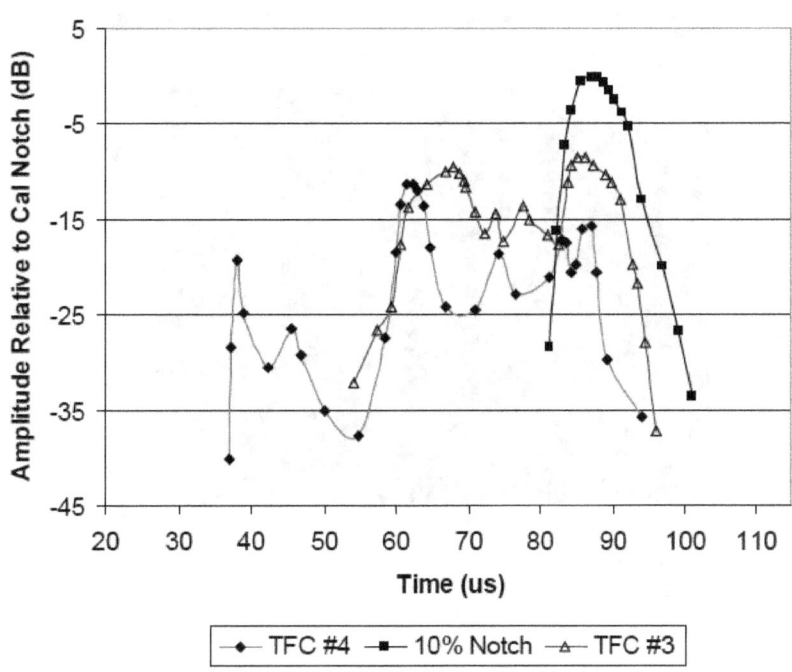

Figure 6-10   Echo-Dynamic Curves for Rough Cracks in Alloy 600 to 45° Shear Waves. Duration of echo-dynamic curves for rough cracks shows some relationship with crack size. TFC 4 is approximately 75% through-wall. TFC 3 is approximately 50% through-wall. The echo-dynamic curve of the 10% notch is shown for reference.

Figure 6-11 shows a scanned image of a thermal fatigue crack in one of the Alloy 600 specimens. Ultrasonic responses are received from the face of the crack. These stacked responses generate the increased duration of the echo-dynamic curve with increasing crack size.

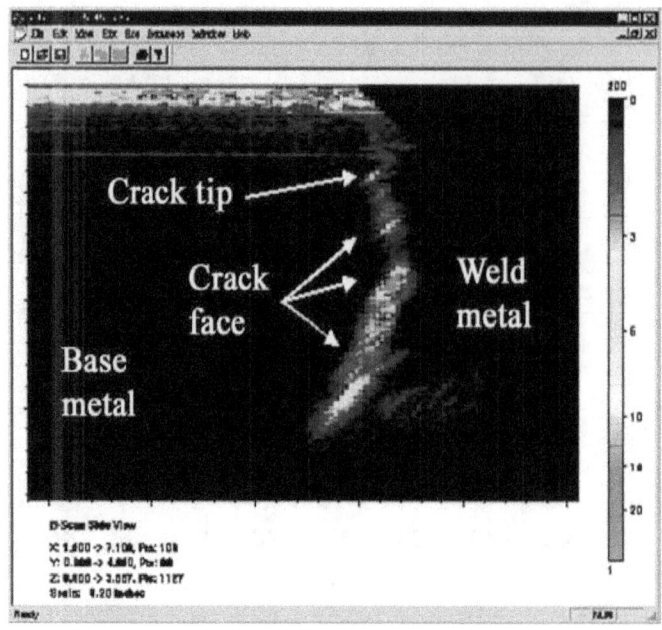

Figure 6-11 Thermal Fatigue Crack in Alloy 600

6.3 Far-Side Through-Weld Metal Access

Far-side access is the second of four access conditions considered in the parametric study. Here the cracks are connected to the surface that is opposite to the scanning surface, and the ultrasonic sound field passes through the weld metal to the cracks.

Figure 6-12 shows the response-time chart from a low-detection performance case, the corner trap response of notches to 2.25 MHz, 45° shear waves through weld-metal. The signals from the notches could be found with careful measurement and a priori knowledge of notch location. However, the signals from all of the notches could not be distinguished from all of the weld-noise interference. The TFC and WFC corner-trap responses could not be measured through weld metal. This is consistent with the 20-dB reduction in response found for notches when inspecting through weld metal as opposed to base metal.

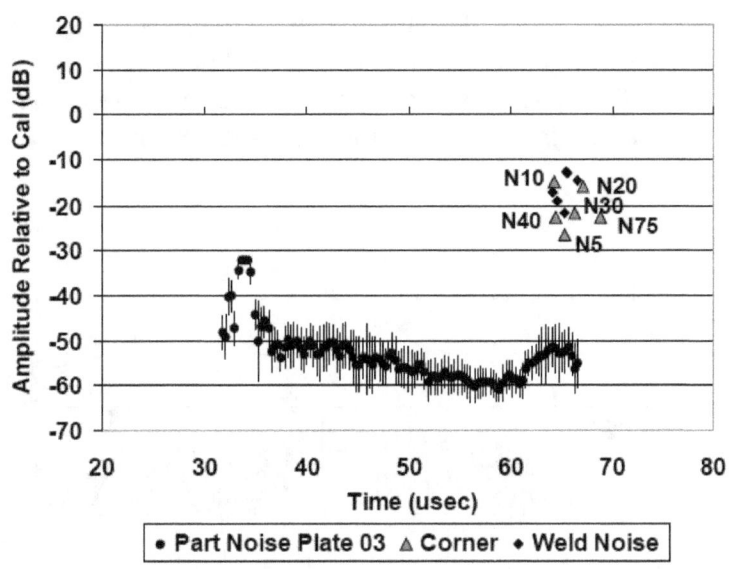

Figure 6-12 **Response of Notches to 45° Shear Waves at 2.25 MHz for Far-Side Through-Weld Metal Access in Stainless Steel**

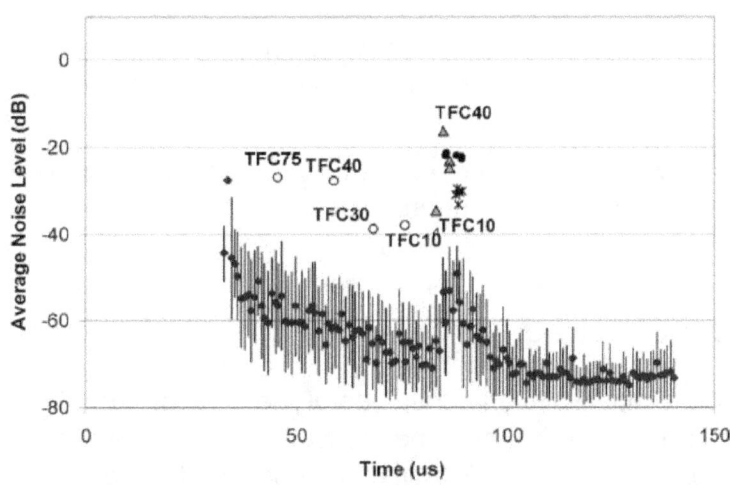

Figure 6-13 **Response of Notches to 45° Shear Waves for Far-Side Through-Weld Metal Access in Alloy 600**

Figure 6-13 shows the response-time chart for notches in Alloy 600 using 45°, 2.25-MHz shear waves for weld metal access. The response from the corner trap signals from the notches varied from −18 dB to −35 dB of the calibration notch. Interference from weld noise is similar to that from stainless steel as shown in Figure 6-12, in that it covers much of the response range for the corner-trap signals. The responses from the tip signals of the notches are shown to be clearly detectable at a reduced level with respect to the case of base metal access.

Figure 6-14 shows the responses for weld metal access to notches using a creeping wave probe in stainless steel. The corner-trap signals could be detected at about 6 dB above the weld noise. The tip signals were found with a greater signal-to-noise ratio than the corner traps because of the absence of fabrication flaws in the interior of the mockup specimens.

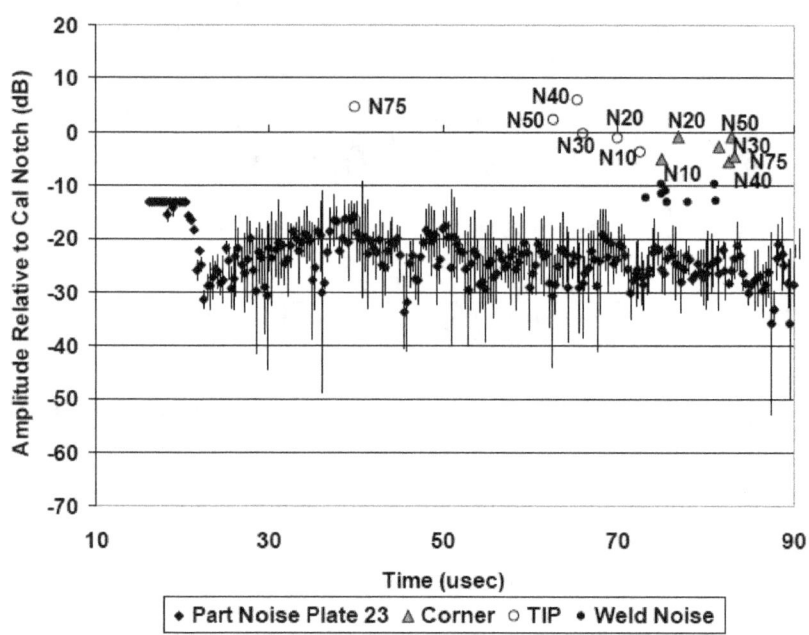

Figure 6-14   Responses of Notches to Creeping Waves at 2 MHz for Far-Side Through-Weld Metal Access in Stainless Steel

6.4   Access to the Cracked Surface

In the core shroud welds, cracks form on either side of the component. This section describes the detection performance for the case in which the ultrasonic transducer can be placed on the surface from which the crack originates. Results for inspection through base metal and weld metal are given.

Figure 6-15 shows the detection performance for a 70°, 2-MHz dual-element probe in the case of access to the cracked surface in stainless steel. The data shown are for notches inspected through the base metal and through the weld metal. The responses were approximately the same for weld metal and base metal access. The responses from the notches did not vary much with the through-wall size of the notch.

Figures 6.16 and 6.17 show the responses for TFCs and WSCs, respectively. The responses from the cracks do not differ from the responses from the notches.

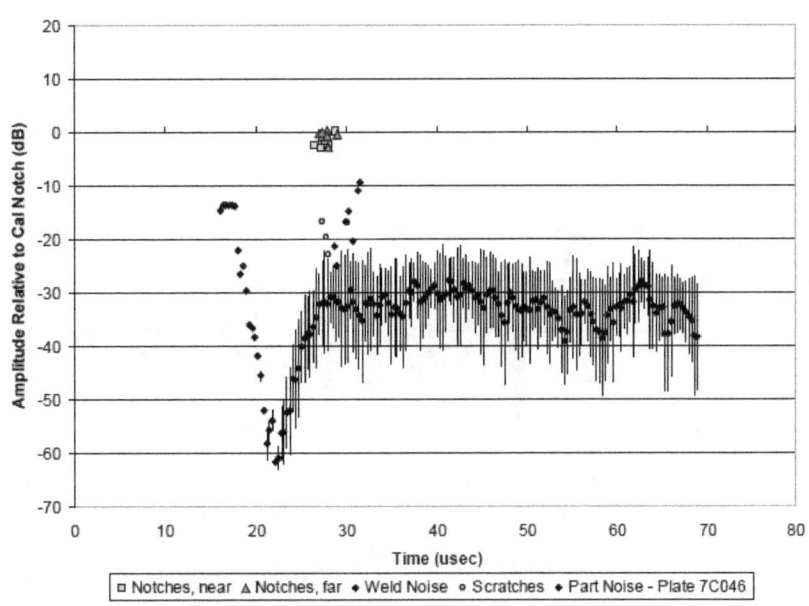

Figure 6-15 Dual-Element Probe, 70° 2-MHz, Detection Results for Notches Connected to Scanning Surface in Stainless Steel

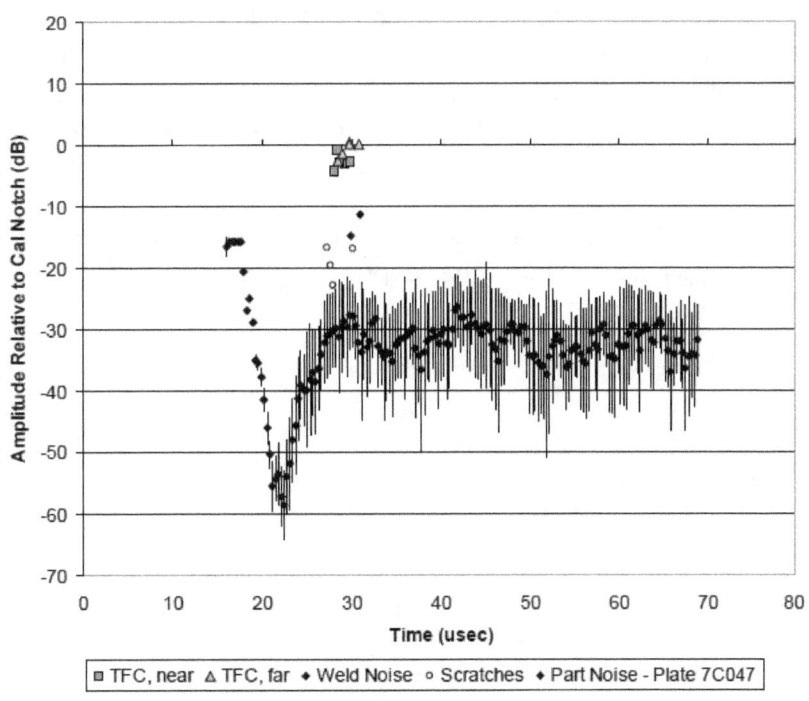

Figure 6-16 Dual-Element Probe, 70° 2-MHz, Detection Results for Thermal Fatigue Cracks Connected to the Scanning Surface in Stainless Steel

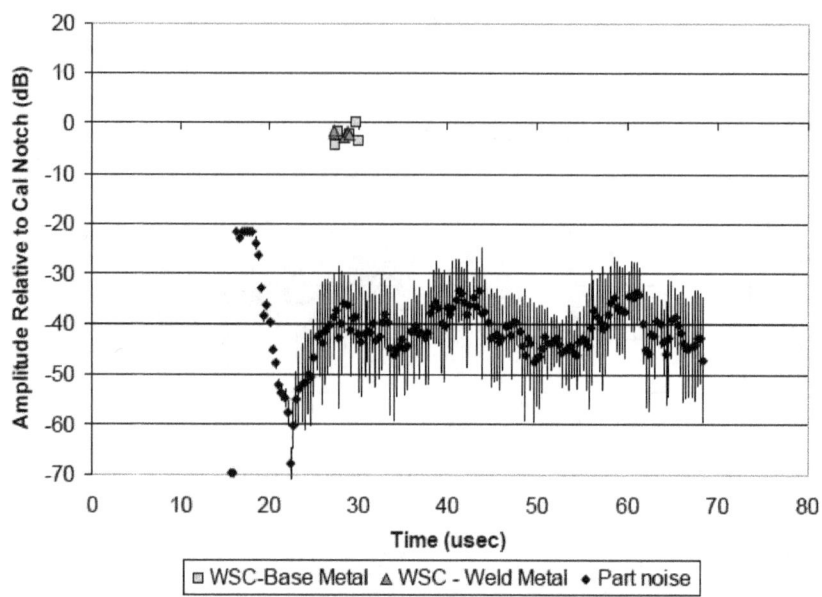

Figure 6-17 Dual-Element Probe, 70° 2-MHz, Detection Results for Weld Solidification Cracks Connected to the Scanning Surface in Stainless Steel

Tip signals were measured for the case of access to the cracked surface using a 60°, 2-MHz dual-element probe. The responses from notch tips are shown for inspection through base metal in Figure 6-18 and through weld metal in Figure 6-19. The responses for inspection through weld metal are 5 dB lower than through base metal but still provide a good signal-to-noise ratio for the defects. Some of the notch tip signals were not detected.

Figures 6-20 and 6-21 show the responses for tip diffraction from TFCs and WSCs. The responses from the crack tips are similar to the responses from the notch tips. Sizing error is described in the next section.

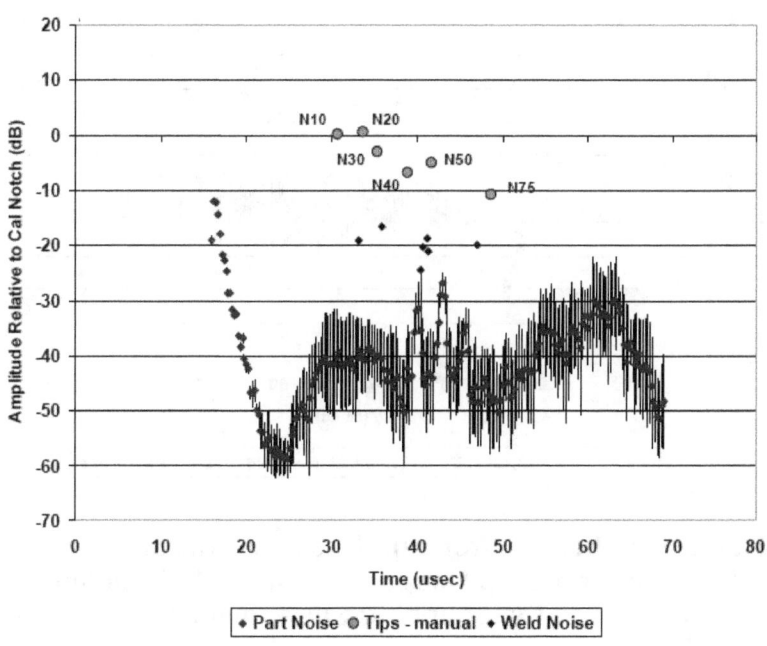

Figure 6-18 Dual-Element Probe, 60° 2-MHz, Notch Tip Responses Through Base Metal for Notches Connected to the Scanning Surface in Stainless Steel

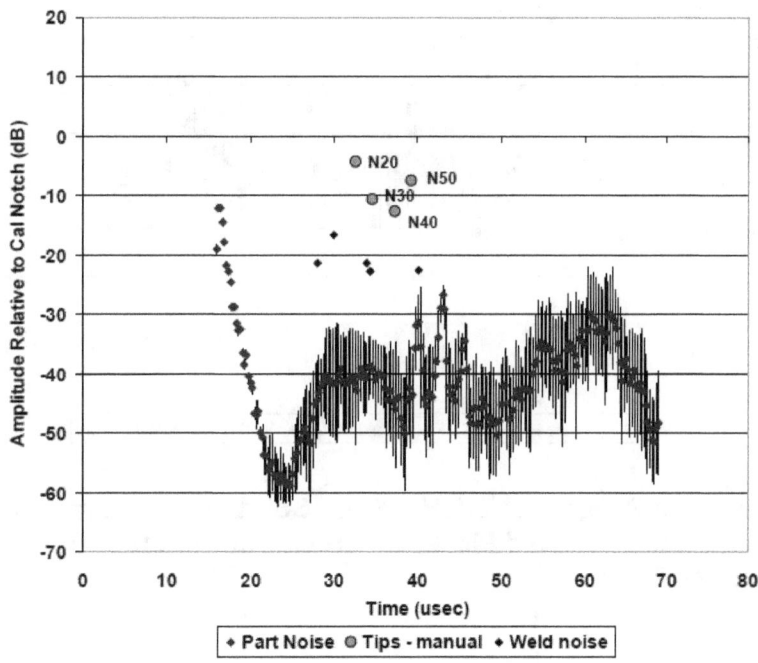

Figure 6-19 Dual-Element Probe, 60° 2-MHz, Notch Tip Responses Through Weld Metal for Notches Connected to the Scanning Surface in Stainless Steel

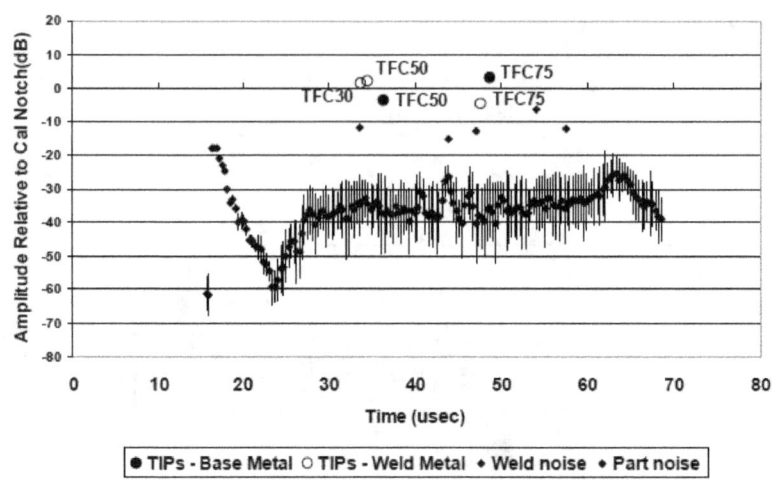

Figure 6-20 Dual-Element Probe, 60° Transmit-Receive Refracted Longitudinal 2-MHz, Thermal Fatigue Crack Tip Responses Through Base and Weld Metal for Cracks Connected to the Scanning Surface in Stainless Steel

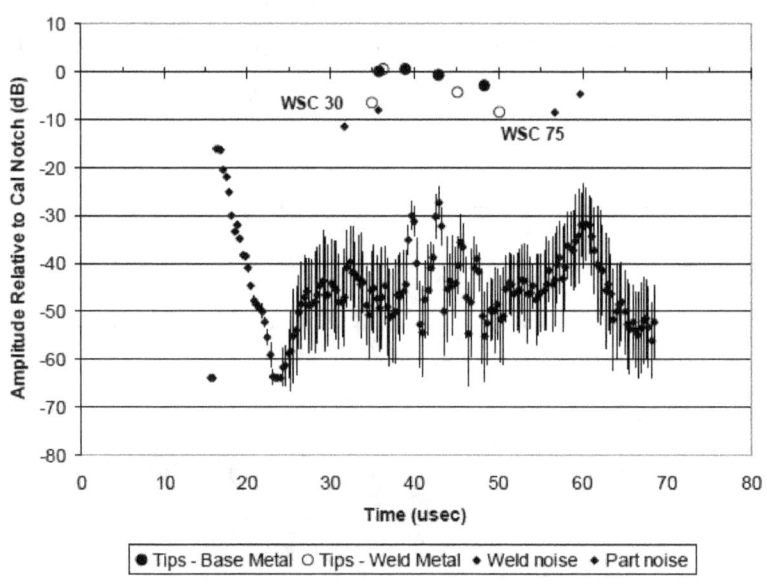

Figure 6-21 Dual-Element Probe, 60° 2-MHz, Weld Solidification Crack Tip Responses Through Base and Weld Metal for Cracks Connected to the Scanning Surface in Stainless Steel

6.5 Summary of Results

This section summarizes the detection and sizing performance provided in the time-response charts from the parametric study. Detection results use the signal-to-noise ratio for the minimum size reflector. Sizing results are given in root mean square sizing error.

Table 6-1 summarizes the detection results for notches that are not connected to the scanning surface. Signal-to-noise ratio for corner-trap responses are listed for detection through base metal and weld metal. The creeping wave probe did not detect the 5% notch when inspecting through base metal. For weld metal access, the creeping wave probe did not detect notches less than 30% through-wall.

Table 6-2 summarizes the detection results for reflectors connected to the scanning surface. All reflectors were detected with a signal-to-noise ratio of 4 to 1.

Sizing results for the case of notches connected to the surface opposite to the ultrasonic transducer are listed in Table 6-3. The 45° 2.25-MHz, shear wave probe had a root mean square error (RMSE) of less than 1 mm when inspecting through base metal. For the case of sizing notches that are connected to the scanning surface, Table 6-4 shows that the RMSE was less than 1 mm for base metal and weld metal access.

The data show that in the case of a high ratio of signal to systematic noise, a parametric study can be used to quantify and assess NDE reliability. Such was possible in the case of reactor internals because of the lack of weld root, counterbore, and geometrical reflectors that made piping inspections difficult. Inspection through austenitic weld metal for cracks that are opposite-surface connected was shown to have low signal to systematic noise performance and the next section shows that a blind test is needed to quantify performance.

Table 6-1   Corner-Trap Signal-to-Noise Ratio for Access Conditions, Ultrasonic Modality, and Notches That Are Opposite-Surface Connected (from the Transducer) in Stainless Steel

Modality	Access	Notch		TFC		WSC	
		SNR	Size	SNR	Size	SNR	Size
45° Shear	Near	2	5%	2	10%	<1	-
(2.25 MHz)	Far	<1	-	<1	-	<1	-
Creeping Wave	Near	2	5%	2	10%	<1	-
(2.0 MHz)	Far	2	30%	<1	-	<1	-

Table 6-2   Detection Signal-to-Noise Ratio for Reflectors Connected to Scanning Surface in Stainless Steel

Modality	Access	Notch		TFC		WSC	
		SNR	Size	SNR	Size	SNR	Size
70° Compression	Near	4	5%	4	10%	4	10%
(2.0 MHz)	Far	4	5%	4	10%	4	10%

Table 6-3 Sizing Error for Notches That Are Opposite-Surface Connected (from the Transducer) in Stainless Steel. No sizing error estimate for thermal fatigue cracks or weld solidification cracks was available at the time of publication. For responses from crack tips, see Figure 6-4 and Figure 6-13.

Modality	Access	RMSE Notches	
		(%)	(mm)
45° Shear	Near	1.3	0.6
(2.25 MHz)	Far	no data	no data
Creeping Wave	Near	6.5	3.3
(2.0 MHz)	Far	9.7	4.9

Table 6-4 Sizing Error for Notches Connected to Scanning Surface in Stainless Steel. No sizing error estimate for thermal fatigue cracks or weld solidification cracks was available at the time of publication. For response from crack tips, see Figures 6-20 and 6-21.

Modality	Access	RMSE Notches	
		(%)	(mm)
60° Compression	Near	1.2	0.6
(2.0 MHz)	Far	0.5	0.3

7 BLIND TEST OF PHASED ARRAY PROCEDURE FOR FAR-SIDE ACCESS

At the EPRI Reactor Pressure Vessel Inspection Conference held in Santa Fe, New Mexico, on June 22–24, 1998, an ISI vendor and PNNL agreed to conduct a blind test evaluation of the ISI vendor's phased array inspection procedure that was developed for the BWR core shroud. The vendor conducted inspections of the PNNL core shroud mockup at the EPRI NDE Center in North Carolina on July 27–31, 1998. PNNL shipped the mockup specimens to the EPRI NDE Center and had two PNNL staff members present during the week of the testing. The principal objective was to measure the inspection performance for the most difficult case of inspection through stainless steel weld metal for cracks connected to the surface opposite from the scanning surface. The secondary objective was to measure the parametric performance of the phased array technique—especially the signal and noise levels.

PNNL laboratory measurements showed that inspecting through weld metal to access cracks connected to the opposite surface is difficult. Consequently, a blind test was conducted of approximately 381 linear centimeters (150 inches) of weld with the cracks on the far side of the weld. Practice specimens were made available to the inspection staff prior to conducting the blind test. The vendor recorded the detection results as start and stop coordinates, and PNNL applied these data to the grading units that were established prior to testing. For each detection, the vendor provided the depth and response of the crack's tip signal.

7.1 Test Description

PNNL shipped 20 mockup assemblies to the EPRI NDE Center. The size and weight of the specimens permitted the manual handling and presentation during the test. Figure 7-1 shows the mockup shipment box sent to the EPRI NDE Center. The specimens were prepared for the blind test by taping all of the grading units to obscure the true state. Tape was also placed over the identifying part numbers provided by the vendor of the specimens to obscure their permanent identity. The blind test was limited to the weld far-side condition. PNNL staff members performed the handling and presentation of the mockup specimens.

Calibration scans were performed for all transducers used in the inspections of the PNNL mockup. Instrument settings were controlled so that, except for receiver gain, they were left unchanged after being established during calibration. Table 7-1 describes the stainless steel calibration specimen provided to the blind test. The mockup specimen, labeled C1, was provided to the vendor before the blind test was conducted. A single calibration reflector, a 10% end-milled notch, was provided. The calibration reflector was used to test the phased array probe and establish system settings. Instrument settings, such as phased array focal laws, were left unchanged after calibration. The blind test required that the only system parameter that could be changed was the overall receiver gain. The receiver gain could be changed to add sensitivity for the inspection of the cracks in the blind test. Practice specimens were provided to determine the needed additional sensitivity.

Figure 7-1 PNNL Mockup at EPRI NDE Center

Table 7-1 PNNL Stainless Steel Calibration Specimen Provided for Blind Test

Specimen Name	Description
C1	10% end milled notch

Practice specimens were inspected with the phased array system. Scans of blank specimens with and without weld crown were made to establish the inspection sensitivity for blank material. A specimen with a thermal fatigue crack, including the true-state data for the crack, was made available. Measurements of the response from the corner trap and crack tip were successfully recorded. A specimen with a weld solidification crack was made available, and measurements of corner trap and crack tip signals were successfully made. Table 7-2 describes the practice specimens provided to the blind test.

Table 7-2 Stainless Steel Practice Specimens Provided for Blind Test

Specimen Name	Description
P1	Blank without weld crown
P2	Blank with weld crown
P3	TFC vertical, TFC tilted
P4	WSC

Note: The thermal fatigue crack and weld solidification crack were approximately 30% through-wall. Exact sizes are withheld so that the cracked specimens can be reused in blind tests.

These practice specimens were inspected after calibration and before the start of the blind test. Two blank specimens were provided to establish the inspection sensitivity for the baseline condition. The blank practice specimen, P1, was presented, and no detectable indications were found. The weld crown on the blank practice specimen, P2, did not produce detectable indications or other difficulties for the phased array technique. A specimen, P3, with two thermal fatigue cracks was presented for far-side inspection (through the weld), and the true-state parameters, location, depth, and length, for the cracks were provided. Measurements of location, depth, and length for the two TFCs in P3 were easily achieved. A practice specimen, P4, was presented with a WSC, and the measurements compared favorably to the true-state information.

A blind test of performance for the inspection of cracks connected to the opposite surface was made. Inspection of 20 assemblies containing cracks adjacent to the weld was conducted with emphasis on far-side access. The assemblies were presented in a predetermined order. Responses from detectable TFCs and WSCs were recorded by the vendor procedure. The start and end of detectable cracks were recorded, based on the vendor procedure, for length sizing. Depth size of detectable cracks was estimated. Figure 7-2 shows the vendor's inspection scanner and the presentation of specimens in the blind test. The inspections with conventional UT (as shown) and phased array ultrasonic probes were conducted using a pipe scanner and not the in-vessel manipulator for inspection of reactor internals. For this study, we were interested in assessing the NDE inspection technology and not the in-vessel manipulator. The conventional probe is shown in the photograph to illustrate the scanning method. However, no evaluation of a conventional UT technique was performed in the blind test. The blind test was conducted solely to provide the detection and sizing performance of the phased array technique.

Figure 7-2 Inspection Scanner and Presentation of Specimens in Blind Test

The blind test had varying amounts of similarity to actual field conditions. Factors that were similar to field conditions were presence of weld crown, a range of surface roughness conditions, coupling, transducer, and instrument settings. The factors that were different from field conditions were the use of artificially implanted flaws, the use of a pipe scanner instead of the actual core shroud scanner, and the orientation of the mockup specimens (lying down) versus a core shroud cylinder.

The core shroud scanner was not used in the blind test. However, because this is a large device, the positioning error, length sizing error, and repeatability are expected to be different from that of a pipe scanner such as the one used in the blind test. Depth sizing error should not be affected by the scanner employed because the transducer is in contact with the component in both cases.

7.2 Detection Performance

Fifty grading units made of type 304 stainless steel were presented in the blind test. Of these grading units, 17 were cracked and 33 were blank. Fourteen of the grading units were correctly characterized as cracked. Three cracks were reported in the 33 blank grading units. These detection results are summarized in Table 7-3.

Table 7-3   Detection Performance for Far-Side Access and Opposite Surface Connected Cracks in Blind Test of Stainless Steel Specimen

Detections (%)	False Calls (%)
82	9

Crack type did not influence detection performance. WSCs and TFCs were detected with equal frequency. Crack tilt did not contribute either. All cracks that were larger than 15 mm (30%) in through-wall extent were detected. Of the three missed cracks, two were more than 5 mm (10%) in through-wall extent and one was more than 25 mm (25%) in through-wall extent. The largest missed crack, more than 25 mm, was a TFC in a specimen with weld crown. For the smaller cracks, one was a TFC and the other a WSC in specimens with weld ground flush.

Three false calls were made, and all of them were in grading units with the weld ground flush. A 10% through-wall crack was called in a grading unit with a weld repair. A 15% through-wall crack was called in simple blank weld metal. A false call was made when an indication in blank material was characterized as a 50% through-wall flaw. Two fabrication flaws, in blank and cracked grading units, were reported, but because their size was reported as small, without through-wall extent, they were not counted as false calls. This false call rate is quite low when compared to the results for far-side access in PNNL's piping inspection round robin, where probability of detection and false call probability were shown to be approximately equal— detection performance not different from those achieved by random guessing (Heasler and Doctor 1996).

7.3 Depth Sizing Performance

Tip-diffracted signals were used to depth size all of the indications in the blind test. Table 7-4 shows the RMSE for the depth size measurements for different crack conditions. For all detected cracks, the RMSE is 9.2 mm of depth. Figure 7-3 shows the depth sizing performance trend line for all cracks. The data show that the technique tends to oversize small cracks and undersize large ones.

Table 7-4 Blind Test Depth Sizing Error for Different Crack Types

Crack Type	RMSE (mm)
All cracks	9.2
Without small cracks (<10 mm)	8.3
TFCs only	10.9
WSCs only	1.6
Without angled cracks	7.7
Without straight cracks	8.4
Weld crown flush	8.2

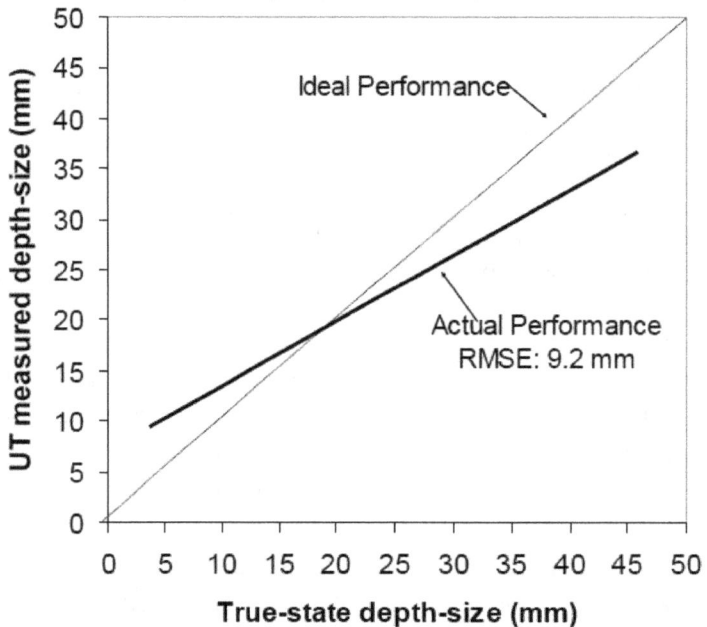

Figure 7-3 Blind Test Depth Sizing Performance for All Detected Cracks

Because there was considerable difficulty detecting and sizing the smaller cracks, the RMSE depth sizing error is recalculated without them. Cracks less than 10 mm (20%) in through-wall extent are considered to be in the small category. The RMSE, at 8.3 mm, did not change much compared to the RMSE for all cracks but Figure 7-4 shows considerable change in the trend line. There is still some tendency to undersize the large cracks but the large effect of the oversizing of small cracks has been removed from the linear trend analysis. The sizing performance on larger cracks shown in this figure is better than Figure 7-3 but there is still some tendency to undersize large cracks.

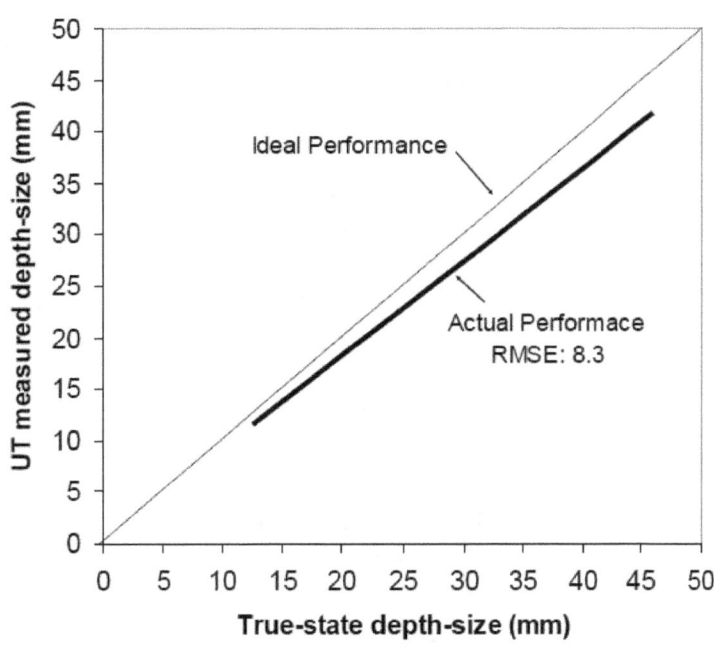

Figure 7-4 Blind Test Depth Sizing Performance Without Small Cracks

Tip diffracted signals were used to depth size all of the indications in the blind test. Table 7-4 shows the RMSE for the depth size measurements for different crack conditions. Of the conditions listed in the table, only the RMSE for WSC cracks is significantly different, acceptably low at 1.6 mm.

The RMSE for depth sizing is better for WSCs than for TFCs when inspecting through austenitic weld metal. The TFCs contributed most to depth sizing error, with an RMSE of 10.9 mm. In an effort to explain this large sizing error, three subsets of the TFC cracks were removed one at a time from the set of all cracks: angled cracks, straight cracks, and cracks with weld crown. Table 7-4 shows that these RMSEs do not differ significantly from the RMSE for all cracks, indicating that none of these single factors contributed most of the sizing error.

Figure 7-5 shows the sizing error dependence on tip response. This scatter diagram shows that the TFC tip signals that contributed most to the sizing error were the ones with the strongest amplitude response. There are a number of explanations for this error: the weld can produce strong responses that cannot be easily distinguished from crack tips, and the anisotropic weld metal refracts the ultrasonic beam, giving a false position for the crack tip. This second explanation requires that the beam steering produces errors of more than 12 mm. Satellite reflectors associated with the TFCs are not the most likely explanation for the depth sizing errors because the amount of weld metal above the TFC should be less than 3 mm.

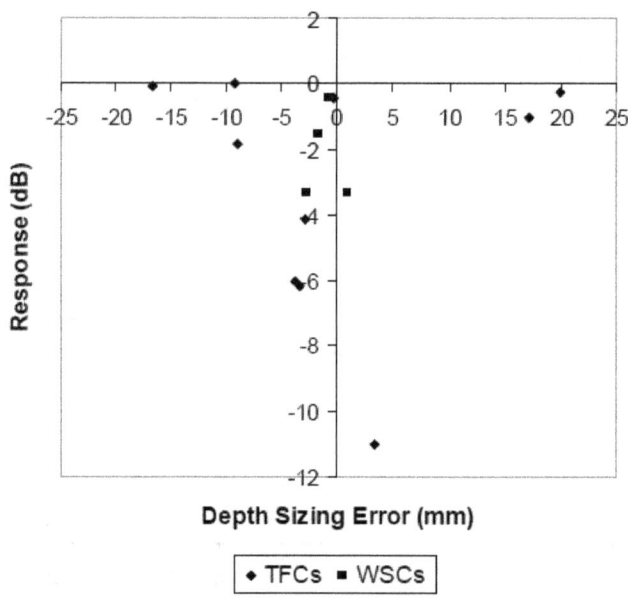

Depth Sizing Error (mm)

◆ TFCs ■ WSCs

Figure 7-5 Blind Test Dependence of Sizing Error on Tip Response

7.4   Length Sizing Performance

Table 7-5 gives the RMSE of 72 mm for length sizing of the 14 cracks detected in the blind test. The length sizing performance for the WSCs was better than that for the TFCs, at an RMSE of 5.8 mm compared to 85 mm.

There were no large undersizing errors, and except for the few large oversizing errors, the crack length was accurately measured. When the oversizing errors are excluded, the RMSE for length sizing is reduced to 10 mm. The length sizing performance was degraded by a number of measurements that exceeded the true length by more than 100 mm, as shown in Figure 7-6. The trend lines with and without the large oversizing errors are given in Figures 7-7 and 7-8, respectively.

Table 7-5 Blind Test Length Sizing Error for Crack by Type

Crack Type	RMSE (mm)
All cracks	72.1
TFCs only	85.3
WSCs only	5.8
Without oversizing (errors < 50 mm)	10.1

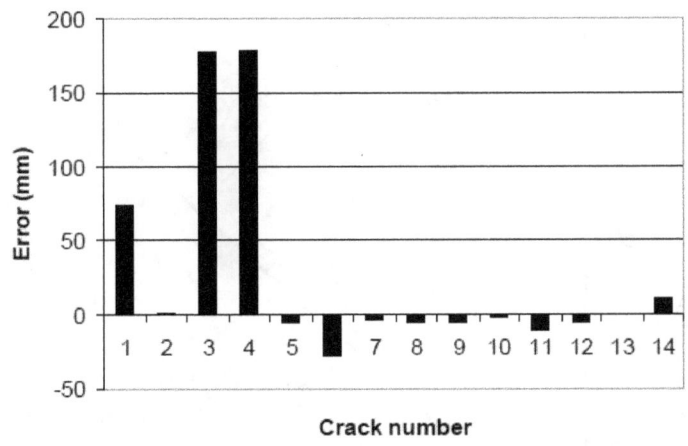

Figure 7-6 Blind Test Length Sizing

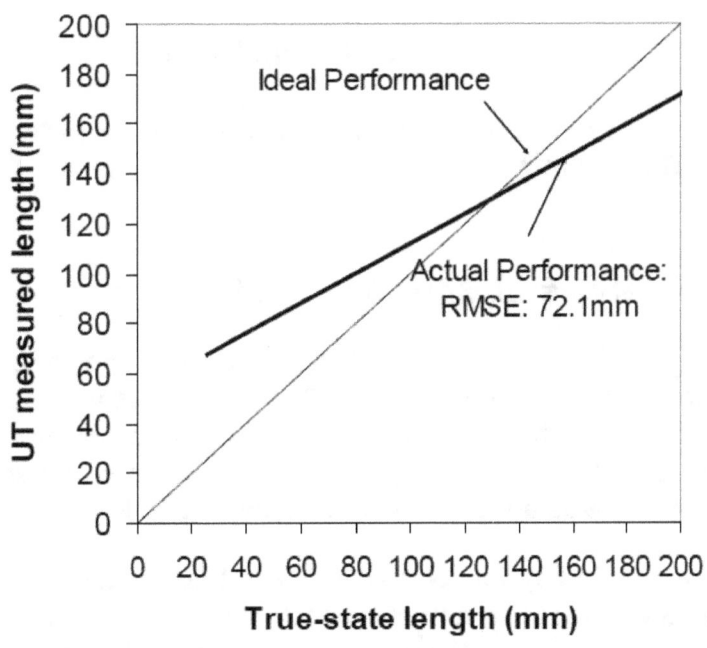

Figure 7-7 Blind Test Length Sizing Performance for All Cracks

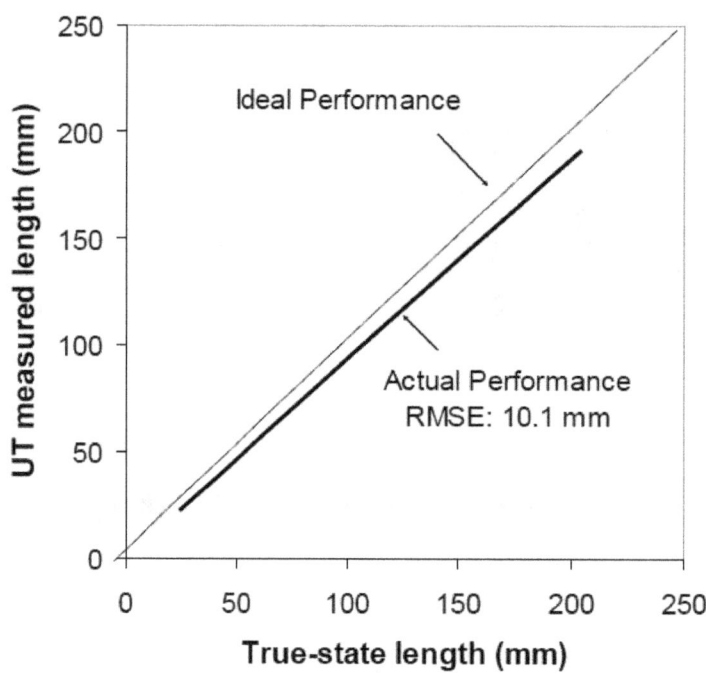

Figure 7-8   Blind Test Length Sizing Performance Without Large Oversizing Errors of Greater Than 50 mm

7.5   Parametric Analysis of Blind Test Data

The mockup's use of a combination of TFCs and WSCs was designed to bracket the range of ultrasonic responses from fully open to very tight SCC. The WSCs are known to give a low response, when inspected through base metal, and as such are considered to be more like tight SCC. The blind test results show that the opposite is the case when inspecting through austenitic weld metal—the WSCs are easier to detect and characterize than the TFCs.

Satellite reflectors were reported by vendor staff in one of the mockup specimens that contained a TFC. Satellite reflectors originate from the flaw implant process and are not meant to include weld noise or part noise but rather only the unintended energy returned from the flaw implant. Vendor staff were concerned about satellite reflectors on only one of the TFC specimens. The TFC specimens do contain a small amount of austenitic weld metal above the crack, and this is a concern because this condition is generally not present in the field unless there are welding flaws or a weld repair at that location. The size of the welded area above the crack is designed to be less than 3 mm. Because the depth sizing error was dominated by errors greater than 7.6 mm and because there was both over- and undersizing, it is reasonable to assume the satellite reflectors did not make a significant contribution to depth sizing error.

Figure 7-9 shows the typical signals from a crack using a phased array system. The azimuthal scan was done electronically (without probe motion) from 30° to 80° with respect to surface normal. The back surface of the part is indicated by the horizontal line at 51 mm (2 in.). The crack signals are above this line. None of the cracks, inspected through the weld by the phased array system, produced a corner trap signal. Signals that crossed the back surface line were determined to be image artifacts. The crack signals in the image are shown between the horizontal 49.87-mm and 18.7-mm lines and the vertical lines at 35.8 mm and 50.9 mm.

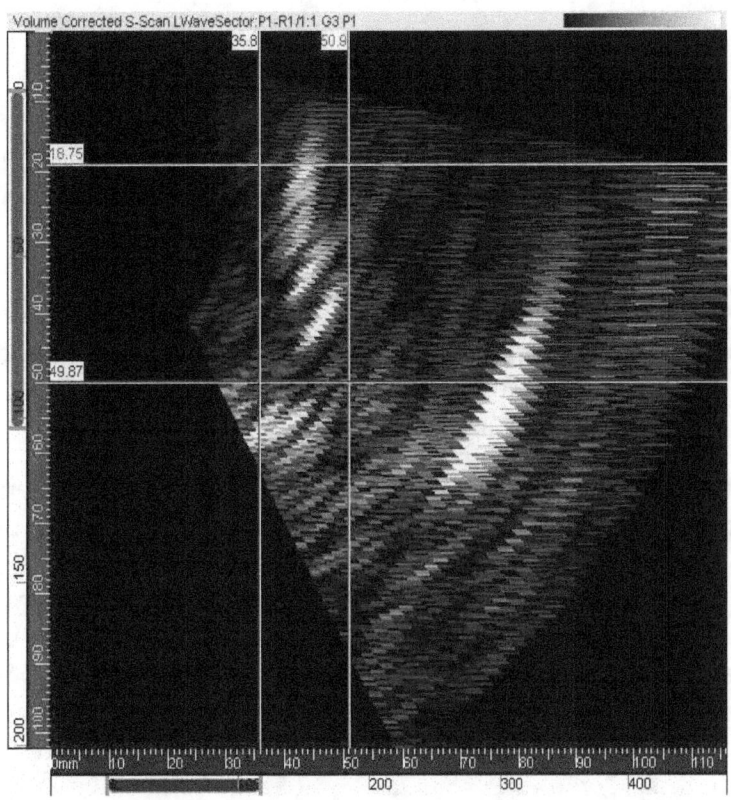

Figure 7-9 Typical Phased Array Image of Crack (units shown in millimeters)

Figure 7-10 shows the relationship between response and crack size in the phased array blind test on the reactor internals mockup. The responses from thermal fatigue cracks and weld solidification cracks are charted by crack size. A logistic curve was fit to the response as a function of the crack size, R(s), defined as

$$R(s) = (1 + exp(\beta_1 - \beta_2 s))^{-1}$$

The fit values are β_1 = 6.0 and β_2 = 0.20 where s, the crack size, is given in percentage through-wall. The reactor internals mockup specimens were 51 mm thick. The RMSE of the logistic function fit to the response data is 9.5% of maximum.

Figure 7-10 shows four cracks with a 10% through-wall size. No response could be found in the inspection data from these four cracks, and these are charted with zero response. One cracked grading unit with a 10% TFC was called cracked in the blind test, but the reflector was a mid-wall fabrication flaw and not the 10% TFC.

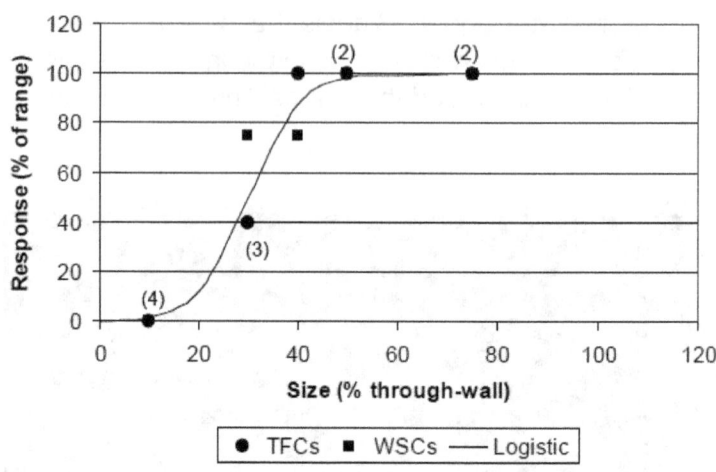

Figure 7-10 Relationship Between Response and Crack Size in Phased Array Blind Test on Stainless Steel in Reactor Internals Mockup

Figure 7-10 shows three 30% through-wall TFCs at 40% of the response range and one 30% WSC at 75% of response range. One of the 30% TFCs was missed in the blind test, but the data review showed that the signal was present. The vendor procedure required that the signal be present in two scans of the phased array, and the missed 30% through-wall crack produced a response in only one scan line. The minimum detected response for the blind test was 40% of maximum. This gives the phased array technique used in the blind test an 8-dB dynamic range in detectable responses.

7.6 Summary

The blind test showed that the vendor phased array inspection procedure could detect and size cracks through austenitic weld metal. The detection performance would meet the intent of a detection performance demonstration test designed to ASME Code Section XI, Appendix VIII Supplement 2, "Qualification Requirements for Wrought Austenitic Piping Welds." The depth sizing RMSE was 9.2 mm for all cracks. The length sizing RMSE was 72 mm for all the detected cracks. The TFCs contributed most of the error to both length and depth sizing measurements.

The depth and length sizing performance would not pass any of the requirements in Appendix VIII. However, the requirements in Appendix VIII are based on the need to detect and characterize small flaws in the components covered by the specific supplements. In the case of the core shroud, the critical flaw size is very large, so it is expected that less stringent requirements might need to be met. Consequently, to put the sizing performance into perspective, further work on assessing requirements for pass/fail criteria is in order.

The parametric study showed that the measurement of responses, using a logarithmic scale, relative to a standard reflector can provide important insights to the sensitivity of the technique to part noise, weld noise, surface scratches, crack tips, and corner-trap responses. The blind test showed that the usefulness of ISI data is increased by such calibration.

8 CONCLUSIONS

The combined use of a parametric study with a blind test can quantify the reliability of the ultrasonic inspection of reactor internals. A mockup, described in Section 5, can be designed for both. The data collected in a parametric study can quantify the performance of ultrasonic in-service inspection techniques and identify areas where performance is relatively low. A blind test can be conducted to further assess the low performance in difficult cases, such as weld far-side access to heat-affected zone cracking on the surface opposite from the one being scanned.

The operating characteristic diagram shows that it is not sufficient to examine probability of detection without also measuring false call probability. Similarly, it is not enough to measure crack response without considering systematic noise from, for example, weld grains. The Programme for the Inspection of Steel Components (Lemaitre 1994) showed that the operating characteristic diagram for procedures in the inspection of stainless steel welds produced large scatter in results by teams that used similar procedures. They concluded that training and performance demonstration testing are needed on realistic specimens. Access to both sides of the weld is needed. Scanning in two different directions, perpendicular to the weld, improved detection performance.

Estimates of response variance are important when quantifying signal-to-noise ratios for cracks and degradation. A PNNL parametric study (Becker et al. 1981) has shown that the standard deviation of crack responses is similar, 6 dB, even when it is derived from different sources. Crack tightness is included in this estimate.

Inspection performance was quantified in this study by identifying the areas in which performance is high and then placing emphasis of subsequent study on the difficult inspection cases. The first part of the study was accomplished by conducting a parametric study, which quantified crack responses and attendant noise sources to show reliable detection and accurate sizing. The second part of the study used a blind test of vendor performance to show detection and sizing statistics for the difficult cases.

8.1 Parametric Study Conclusions

The parametric study of UT-ISI of reactor internals quantified the performance of industry-type techniques and identified the areas in which reliability is relatively low. To accomplish this, the parametric study established the amplitude and pattern of noise from different sources, determined the effects of weld and crack type on signal strength, and established the accuracy of sizing methods.

Results of the parametric study showed that access to the cracked surface strongly affected detection and sizing. In the case of inspection of weld heat-affected zones for cracking, reliable detection and accurate sizing could be accomplished when the ultrasonic transducer could be placed on the cracked surface. Furthermore, reliable detection and accurate sizing could be achieved for the near-side access condition to cracks that originate from the surface opposite to the scanning surface. The parametric study showed that for weld far-side access to cracks, detection and sizing performance was low.

Inspection through austenitic weld metal was a case in which performance was relatively low. Vertically polarized ultrasonic shear waves often failed to detect side-drilled holes through austenitic weld metal. Ultrasonic compression (longitudinal) waves sometimes detected side-drilled holes through the weld metal, but performance was greatly reduced from the base-metal case. The performance of creeping-wave probes, when inspecting through austenitic weld metal, was better than that for the other ultrasonic modalities evaluated in this study.

An analysis of the differences in signal strengths between weld solidification cracks and thermal fatigue cracks showed similar and dissimilar features. Similar UT responses were received from both crack types when the cracks were connected to the scanning surface. For the inspection of cracks connected to the surface opposite the scanning surface, the responses were 18 dB weaker for the WSC with base metal access. For inspection through weld metal, some WSC responses were not detectable. Southwest Research Institute research showed that its WSCs are 60% lower in response than real SCC (Watson and Edwards 1996). Using this estimate, our data showed that TFCs were higher in response than real SCC.

For the TFCs, the cracking process uses a repair weld to complete the flaw implant. It is important to know if the repair weld betrays the presence of the flaw. Our data showed that the implant repair weld does not produce a detectable response when probes typically used by industry are used. As a consequence, we do not believe the TFC weld repair to be a significant concern.

Weld type—submerged vs. shielded metal arc—is less important than weld profile, with shielded metal arc weld being slightly more difficult to penetrate with ultrasound. Core shroud welds are made by submerged metal arc, but because shielded metal arc welds are easier to make and more difficult to inspect, the shielded metal arc process was used as a conservative substitute.

8.2   Blind Test of Phased Array Procedure

A blind test was shown to be needed for measuring UT-ISI performance in those cases where the signal-to-noise levels are relatively low—weld far-side access. Fifty grading units of type 304 stainless steel were presented in the blind test. Of the 50 grading units, 17 were cracked and 33 were blank. Fourteen of the grading units were correctly characterized as cracked. Three cracks were reported in the 33 blank grading units. This detection performance would pass ASME Code Section XI, Appendix VIII Supplement 2, "Qualification Requirements for Wrought Austenitic Piping Welds." The root mean square depth sizing error for the blind test was 9.2 mm and would not pass.

A re-analysis of the signals in the vendor's phased array data revealed that no response was received from cracks less than 30% through-wall for far-side access. The absence of corner-trap responses explains, in part, the size dependence of crack detection. The signals in the phased array images originated from the rough face of the cracks and from the crack tips. Sizing errors originated from difficulty distinguishing the weak responses from the crack tips from the strong responses produced by the crack face and those signals caused by weld fabrication flaws.

Crack type did not influence detection performance. WSCs and TFCs were detected with equal frequency. Crack tilt did not contribute, either. All cracks larger than 15 mm (30%) in through-wall extent were detected.

Three false calls were made, and all of them were in grading units with the weld ground flush. A 10% through-wall crack was called in a grading unit with a weld repair. A 15% through-wall crack was called in a simple blank weldment. A false call was made when an indication in blank material was characterized as a 50% through-wall welding (fabrication) flaw. This false call rate is quite low when compared to the results from PNNL's piping inspection round robin for far-side access of austenitic welds where probability of detection and false call probability were shown to be approximately equal (detection performance not measurably different from guessing).

A combination of TFCs and WSCs in the mockup was designed to bracket the range of ultrasonic responses from fully open to very tight SCC. The WSCs are known to give a low response when inspected through base metal and, as such, are considered to be more like tight SCC. The blind test results show that the opposite is the case when inspecting through austenitic weld metal: the WSCs are easier to detect and characterize than the TFCs.

The data show three 30% through-wall TFCs at 40% of the response range and one 30% WSC at 75% of the response range. One of the 30% TFCs was missed in the blind test, but the data review showed that the signal was present. The minimum detected response for the blind test was 40% of maximum. This gives the phased array technique used in the blind test an 8-dB dynamic range in detectable responses.

9 RECOMMENDATIONS

In this section, recommendations are made where enhancements can be achieved: imaging improvements, such as reduction of noise variance and hybrid imaging systems, and application of the findings of this work to the difficult case of the far-side inspection of austenitic piping welds. Signal-to-noise improvements can be made by reducing systematic noise and its variance from microstructure and surface geometry. Weld microstructure imaging is interesting for the early detection of degradation. Application of systematic noise analysis to piping welds is a significant challenge.

Image quality can be improved through signal-to-noise improvements. The response and variance from weld microstructure can be reduced through spatial averaging. Small indications from weld noise can be suppressed by spatial-frequency filtering. Calibrated responses are useful for understanding variances and their effect on decision making.

Decision thresholds have been an important part of data analysis (Berens 1988). Application of decision thresholds on calibrated ultrasonic responses for flaw detection has been difficult, and current practice is to analyze inspection data to within some multiple of the noise level, typically +6 dB of that level. Still, the significance of calibrated ultrasonic response is great when expressed on a logarithmic scale. As discussed in Section 4, a standard deviation of 6 dB can be expected in calibrated responses from SCC in welded components.

Imaging of weld microstructure could be investigated. New image reconstruction algorithms can improve characterization of the shape and orientation of reflectors in welded assemblies. Sensitive image techniques that image weld grains can be useful for addressing smooth cracks and tight cracks.

Imaging systems can fuse multiple data streams into images of weld condition. A hybrid system, based on phased arrays and creeping waves, might be interesting for data fusion imaging through weld metal.

The successful part of this work on reactor internals should be extended to austenitic piping welds. The ultrasonic in-service inspection of wrought austenitic, cast austenitic, and dissimilar metal welds is more difficult than the inspection of the welds in the BWR core shroud. First, the inside and outside surfaces of pipes complicate ultrasonic inspections (Morris and Becker 1982). Second, the counterbore and weld root on the inside of the pipes produce ultrasonic reflections that are difficult to distinguish from degradation. Imaging of the weld microstructure and the shape and orientation of geometrical conditions could be developed. Optimizing the detection algorithms for application to ultrasonic inspection of piping weldments could be investigated.

10 REFERENCES

10 CFR 50, Appendix B, "Quality Assurance Criteria for Nuclear Power Plants and Fuel Reprocessing Plants." *Code of Federal Regulations*, U.S. Nuclear Regulatory Commission.

ASME. 2004. "Rules for Inservice Inspection of Nuclear Power Plant Components, Section XI." In *ASME Boiler and Pressure Vessel Code – An International Code*. American Society of Mechanical Engineers, New York.

Baker A, LR Fox and RG McClelland. 1994. "Jet Pump Beam Ultrasonic Testing at the Susquehanna Steam Electric Station." Presented at *EPRI Vessel & Internals Inspection Conference*, July 11–15, 1994, San Antonio, Texas.

Baker A and J Van Hoomissen. 1994. "BWR Core Shroud Inspection Program." Presented at *EPRI Vessel & Internals Inspection Conference*, July 11–15, 1994, San Antonio, Texas.

Becker FL, SR Doctor, PG Heasler, CJ Morris, SG Pitman, GP Selby and FA Simonen. 1981. *Integration of NDE Reliability and Fracture Mechanics, Phase 1 Report*. NUREG/CR-1696, PNL-3469 Vol. 1, U.S. Nuclear Regulatory Commission, Washington, D.C.

Berens AP. 1988. "NDE Reliability Data Analysis." In *Metals Handbook, Volume 17: Nondestructive Evaluation and Quality Control*, pp. 689–701. ASM International, Materials Park, Ohio.

Bertz S, E Black and J Langdon. 1994. "Reactor Core Shroud Inspections at Brunswick Unit 1." Presented at *EPRI Vessel & Internals Inspection Conference*, July 11–15, 1994, San Antonio, Texas.

BWRVIP-03. 1995. *Reactor Pressure Vessel and Internals Examination Guidelines*. TR-105696, Electric Power Research Institute, Boiling Water Reactor Owners Group's Vessel and Internals Project, Palo Alto, California.

BWRVIP-15. 1996. *Configuration of Safety-Related BWR Reactor Internals*. TR-106368, Electric Power Research Institute, Boiling Water Reactor Owners Group's Vessel and Internals Project, Palo Alto, California.

BWRVIP-18. 1996. *BWR Core Spray Internals Inspection and Flaw Evaluation Guidelines*. TR-106740, Electric Power Research Institute, Boiling Water Reactor Owners Group's Vessel and Internals Project, Palo Alto, California.

BWRVIP-25. 1996. *BWR Core Plate Inspection and Flaw Evaluation Guidelines*. TR-107284, Electric Power Research Institute, Boiling Water Reactor Owners Group's Vessel and Internals Project, Palo Alto, California.

BWRVIP-26. 1996. *BWR Top Guide Inspection and Flaw Evaluation Guidelines*. TR-107285, Electric Power Research Institute, Boiling Water Reactor Owners Group's Vessel and Internals Project, Palo Alto, California.

BWRVIP-27. 1997. *BWR Liquid Standby Control System/Core Plate ΔP Inspection and Flaw Evaluation Guidelines.* TR-107286, Electric Power Research Institute, Boiling Water Reactor Owners Group's Vessel and Internals Project, Palo Alto, California.

BWRVIP-38. 1997. *BWR Shroud Support Inspection and Flaw Evaluation Guidelines.* TR-108823, Electric Power Research Institute, Boiling Water Reactor Owners Group's Vessel and Internals Project, Palo Alto, California.

BWRVIP-41. 1997. *BWR Jet Pump Assembly Inspection and Flaw Evaluation Guidelines.* TR-108728, Electric Power Research Institute, Boiling Water Reactor Owners Group's Vessel and Internals Project, Palo Alto, California.

BWRVIP-42-A. 2005. *BWR LPCI Coupling Inspection and Flaw Evaluation Guidelines.* TR-1011470, Electric Power Research Institute, Boiling Water Reactor Owners Group's Vessel and Internals Project, Palo Alto, California.

BWRVIP-47. 1997. *BWR Lower Plenum Inspection and Flaw Evaluation Guidelines.* TR-108727, Electric Power Research Institute, Boiling Water Reactor Owners Group's Vessel and Internals Project, Palo Alto, California.

BWRVIP-48. 1998. *Vessel ID Attachment Weld Inspection and Flaw Evaluation Guidelines.* TR-108724, Electric Power Research Institute, Boiling Water Reactor Owners Group's Vessel and Internals Project, Palo Alto, California.

BWRVIP-49-A. 2002. *Instrument Penetration Inspection and Flaw Evaluation Guidelines.* TR-1006602, Electric Power Research Institute, Boiling Water Reactor Owners Group's Vessel and Internals Project, Palo Alto, California.

BWRVIP-76. 1999. *BWR Core Shroud Inspection and Flaw Evaluation Guidelines.* TR-114232, Electric Power Research Institute, Boiling Water Reactor Owners Group's Vessel and Internals Project, Palo Alto, California.

BWRVIP-139-A. 2009. *Steam Dryer Inspection and Flaw Evaluation Guidelines.* TR-1018794, Electric Power Research Institute, Boiling Water Reactor Owners Group's Vessel and Internals Project, Palo Alto, California.

BWRVIP-180. 2007. *Access Hole Inspection and Flaw Evaluation Guidelines.* TR-1013402, Electric Power Research Institute, Boiling Water Reactor Owners Group's Vessel and Internals Project, Palo Alto, California.

BWRVIP-183. 2007. *Top Guide Grid Beam Inspection and Flaw Evaluation Guidelines.* TR-1013401, Electric Power Research Institute, Boiling Water Reactor Owners Group's Vessel and Internals Project, Palo Alto, California.

BWRVIP. 1997. *BWR Vessel and Internals Project, Vessel Internals Inspections Summaries, April 1997.* TR-105696, Electric Power Research Institute, Boiling Water Reactor Owners Group's Vessel and Internals Project, Palo Alto, California.

Davis JB and SA Huntington. 1998. "Automated Tooling for the Inspection of BWR Internals Components." Presented at *3rd EPRI Reactor Pressure Vessel Inspection Conference,* June 22–24, 1998, Santa Fe, New Mexico.

Doctor SR and FL Becker. 2002. "Two Decades of Improvement in Austenitic Stainless Steel Piping ISI." In *Joint EC-IAEA Technical Meeting on Improvements in In-Service Inspection Effectiveness.* November 19–21, 2002, Petten, The Netherlands. International Atomic Energy Agency, Department of Nuclear Energy, Division of Nuclear Power, and European Commission, Joint Research Centre, Institute of Energy, Petten, The Netherlands.

Fisher E and M Tagliamonte. 1994. "New Developments in Advanced Technologies for In-Vessel Exams." Presented at *EPRI Vessel & Internals Inspection Conference*, July 11–15, 1994, San Antonio, Texas.

Forli O. 1979a. "Comparison of Radiography and Ultrasonic Testing." In *Second Nordiske NDT Symposium.* May 21–23, 1979, Kobenhaum, Denmark.

Forli O. 1979b. "Reliability of Ultrasonic and Radiographic Testing." In *Ninth World Conference on Nondestructive Testing.* November 19–23, 1979, Melbourne, Australia.

GE Nuclear Energy. 1996. "Jet Pump Riser Pipe Cracking." GE Nuclear Energy, Inc., Wilmington, North Carolina

Good MS and LG Van Fleet. 1986. *Status of Activities for Inspecting Weld Overlaid Pipe Joints.* NUREG/CR-4484, PNNL-5729, U.S. Nuclear Regulatory Commission, Washington, D.C.

Hacker MG, MP Levesque, GE Whitman, TR Crippes and J Schanen. 1998. "Ultrasonic Examination of Jet Pump Mixer, Diffuser, and Adapter Welds." Presented at *3rd EPRI Reactor Pressure Vessel Inspection Conference*, June 22–24, 1998, Santa Fe, New Mexico.

Hayden JJ. 1998. "Phased Array Ultrasonic Examination of BWR Core Shroud Assembly Welds." Presented at *3rd EPRI Reactor Pressure Vessel Inspection Conference*, June 22–24, 1998, Santa Fe, New Mexico.

Heasler PG, DJ Bates, TT Taylor and SR Doctor. 1986. *Performance Demonstration Tests for Detection of Intergranular Stress Corrosion Cracking.* NUREG/CR-4464, PNL-5705, U.S. Nuclear Regulatory Commission, Washington, D.C.

Heasler PG and SR Doctor. 1996. *Piping Inspection Round Robin.* NUREG/CR-5068, PNL-10475, U.S. Nuclear Regulatory Commission, Washington, D.C.

Heasler PG, TT Taylor, JC Spanner, SR Doctor and JD Deffenbaugh. 1990. *Ultrasonic Inspection Reliability for Intergranular Stress Corrosion Cracks: A Round Robin Study of the Effects of Personnel, Procedures, Equipment and Crack Characteristics.* NUREG/CR-4908, PNL-6179, U.S. Nuclear Regulatory Commission, Washington, D.C.

Herrera ML and PP Stancavage. 1988. "BWR Internals Life Assurance." In *American Nuclear Society/European Nuclear Society/Atomic Energy Society of Japan Topical Meeting on Nuclear Power Plant Life Extension*, pp. 189–196. July 31–August 3, 1988, Snowbird, Utah. American Nuclear Society, La Grange, Illinois.

Lemaitre P. 1994. *Report on the Evaluation of the Inspection Results of the Wrought-to-Wrought PISC III Assemblies No. 31, 32, 33, 34, 35 and 36.* PISC III Report No. 33, Programme for Inspection of Steel Components, Joint Research Centre, EEC, Petten, The Netherlands.

Lemaitre P, TD Koble and SR Doctor. 1996. "Summary of the PISC Round Robin Results on Wrought and Cast Austenitic Steel Weldments, Part I: Wrought-to-Wrought Capability Study." *International Journal of Pressure Vessels and Piping* 69(1):5–19.

Lilley JR. 1994. "Developments in Vessel Weld Inspection using TOFD." Presented at *EPRI Vessel & Internals Inspection Conference*, July 11–15, 1994, San Antonio, Texas.

MacDonald DE. 1994. "Effectiveness of BWR Shroud Support Plate Access Hole Cover Examinations." Presented at *EPRI Vessel & Internals Inspection Conference*, July 11–15, 1994, San Antonio, Texas.

Morris CJ and FL Becker. 1982. *State-of-Practice Review of Ultrasonic In-service Inspection of Class I System Piping in Commercial Nuclear Power Plants.* NUREG/CR-2468, PNL-4026, U.S. Nuclear Regulatory Commission, Washington, D.C.

Shah VK and PE MacDonald. 1993. *Aging and Life Extension of Major Light Water Reactor Components.* Elsevier Science Publishers B.V., Amsterdam, The Netherlands.

Silk MG. 1978. *Estimates of the Magnitude of Some of the Basic Sources of Error in Ultrasonic Defect Sizing.* AERE-R-9023, Atomic Energy Research Establishment, Harwell, Oxfordshire, United Kingdom.

Swets JA. 1983. "Assessment of NDT Systems - Part 1: The Relationship of True and False Detection." *Materials Evaluation* 41:1294–1303.

Swets JA and RM Pickett. 1982. *Evaluation of Diagnostic Systems: Methods from Signal Detection Theory.* Academic Press, New York.

USNRC. 1980. *BWR Jet Pump Assembly Failure.* IE Bulletin No. 80-07 (April 4, 1980) and Supplement 1 (May 13, 1980), U.S. Nuclear Regulatory Commission, Washington, D.C.

USNRC Technical Training Center. *Systems Manual, Boiling Water Reactors, BWR/6 Design.* U.S. Nuclear Regulatory Commission, Washington, D.C.

Watson PD and RL Edwards. 1996. "Fabrication of Test Specimens Simulating IGSCC for Demonstration and Inspection Technology Evaluation." In *14th International Conference on NDE in the Nuclear Pressure Vessel Industries*, pp. 165–168. September 24–26, 1996, Stockholm, Sweden. ASM International, Materials Park, Ohio.

NRC FORM 335
(12-2010)
NRCMD 3.7

U.S. NUCLEAR REGULATORY COMMISSION

BIBLIOGRAPHIC DATA SHEET

(See instructions on the reverse)

1. REPORT NUMBER
(Assigned by NRC, Add Vol., Supp., Rev., and Addendum Numbers, if any.)

NUREG/CR-7159

2. TITLE AND SUBTITLE

Reliability of Ultrasonic In-Service Inspection of Welds in Reactor Internals of Boiling Water Reactors

3. DATE REPORT PUBLISHED

MONTH	YEAR
April	2013

4. FIN OR GRANT NUMBER
Y6604, N6398

5. AUTHOR(S)
G.J. Schuster, S.L. Crawford, A.A. Diaz, P.G. Heasler, and S.R. Doctor

6. TYPE OF REPORT

Technical

7. PERIOD COVERED (Inclusive Dates)

8. PERFORMING ORGANIZATION - NAME AND ADDRESS (If NRC, provide Division, Office or Region, U. S. Nuclear Regulatory Commission, and mailing address; if contractor, provide name and mailing address.)
Michael T. Anderson, PNNL Project Manager
Mail Stop K5-26, PO Box 999
Pacific Northwest National Laboratory
Richland, WA 99352

9. SPONSORING ORGANIZATION - NAME AND ADDRESS (If NRC, type "Same as above", if contractor, provide NRC Division, Office or Region, U. S. Nuclear Regulatory Commission, and mailing address.)
Division of Engineering
Office of Nuclear Regulatory Research
U.S. Nuclear Regulatory Commission
Washington, DC 20555-0001

10. SUPPLEMENTARY NOTES
W.E. Norris, Project Manager

11. ABSTRACT (200 words or less)
Instances of stress corrosion cracking in reactor pressure vessel internal components have been found, especially in the boiling water reactor (BWR) core shroud. Results from in-service inspection are an important aspect of integrity evaluations. One of the major goals of the work described in this report is to quantify the crack detection reliability and sizing error of ultrasonic inspection methods for reactor internals. A mockup is described, along with its application to the assessment on nondestructive evaluation reliability for in-service inspection of reactor internals. The mockup of a BWR core shroud includes cracks in some of the 40 welded assemblies selected to represent field conditions. The selected material and geometry include most conditions and alloys used in the core shroud and its support structure. This report provides an overview of the work being performed and focuses on the parametric study results relating ultrasonic response and its variance to inspection effectiveness. Results of a blind test of an in-service inspection vendor's phased array technique are also provided in this report.

12. KEY WORDS/DESCRIPTORS (List words or phrases that will assist researchers in locating the report.)
boiling water reactors, BWR
ultrasonic testing, UT
in-service inspection, ISI
BWR reactor internal welds

13. AVAILABILITY STATEMENT
unlimited

14. SECURITY CLASSIFICATION
(This Page)
unclassified

(This Report)
unclassified

15. NUMBER OF PAGES

16. PRICE

Printed
on recycled
paper

Federal Recycling Program

UNITED STATES
NUCLEAR REGULATORY COMMISSION
WASHINGTON, DC 20555-0001

OFFICIAL BUSINESS

NUREG/CR-7159

Reliability of Ultrasonic In-Service Inspection of Welds in Reactor Internals of Boiling Water Reactors

April 2013

www.ingramcontent.com/pod-product-compliance
Lightning Source LLC
Chambersburg PA
CBHW080304180526
45167CB00006B/2668